Physics against Cancer

How the Paul Scherrer Institute pioneered modern proton therapy

By Damien Weber and Simon Crompton

25 November 1996 saw a world first for the Paul Scherrer Institute's Center for Proton Therapy when a cancer patient was treated using the spot scanning technique for the first time. This book was commissioned by PSI to commemorate the 25th anniversary.

Bibliographic information published by the Deutsche Nationalbibliothek
The Deutsche Nationalbibliothek lists this publication in the Deutsche Nationalbibliografie; detailed bibliographic data are available in the Internet at http://dnb.dnb.de.

ISBN 978-3-7281-4136-1

www.vdf.ethz.ch
verlag@vdf.ethz.ch

© 2023, vdf Hochschulverlag AG an der ETH Zürich

All rights reserved. Nothing from this publication may be reproduced, stored in computerised systems or published in any form or in any manner, including electronic, mechanical, reprographic or photographic, without prior written permission from the publisher.

Dedication

This book is dedicated to all the patients and families whose difficult journeys with cancer have included a visit to the Center for Proton Therapy at the Paul Scherrer Institute. Their experiences, tolerance and open-mindedness have been at the centre of our growing understanding of the power of particles to counter cancer. The work here has been with them and for them.

The contributors

This book tells the story of proton therapy at the Paul Scherrer Institute (PSI). Many outstanding people at PSI have used their remarkable skills to move modern cancer treatment forward. With this book, we want to provide a public profile for their work. Inevitably, it is only possible to focus on some of the individuals involved in the story, as representatives of a far larger group of people. It is important to remember that modern research can only be successful if undertaken by a team, and every achievement recounted here was the result of collaboration and information sharing.

The book was written by health and science writer Simon Crompton, based on research, interviews and written accounts from past and present staff at the Paul Scherrer Institute. It was produced in close collaboration with Damien Weber, Head and Chairman of the Center for Proton Therapy at PSI.

Simon Crompton talked to the following people to research the book. The authors owe them a debt of gratitude for their help and time, which has been instrumental and invaluable in producing this book:

Hans Blattmann	Martin Jermann
Emmanuel Egger	Barbara Kaser-Hotz
Carl von Essen	Tony Lomax
Rita Feurer	David Meer
Martina Frei	Eros Pedroni
Gudrun Goitein	Ann Schalenbourg
Richard Greiner	Markus Weiss
Martin Grossmann	Leonidas Zografos

Photographs were researched by Martin Grossmann and Ulrike Kliebsch. Project management and draft-reading support were provided by Ulrike Kliebsch and Dagmar Baroke. Martin Grossmann, Tony Lomax and David Meer provided scientific advice.

A note for readers

The words of contributors gathered from interviews and written accounts specifically for this book are reproduced throughout in italic script. Words quoted from other written accounts – for example, from reports and journal papers – are printed in plain text within quotation marks.

Table of Contents

The contributors 5
National foreword 8
International foreword 10
Introduction: about this story 12
Some important background: radiation therapy 15
A timeline of particle therapy at the Paul Scherrer Institute 21
Chapter 1 1968–1977 Beginnings: the piotron 23
Chapter 2 1977–1983 Treating in three dimensions 38
Chapter 3 1982–1992 Protons for eye tumours 54
Chapter 4 1984–1992 From pions to protons 75
Chapter 5 1993–1996 A revolution in cancer patient treatment 99
Chapter 6 1997–2003 Project PROSCAN 127
Chapter 7 2004–2008 A new lifeline for children 150
Chapter 8 2009–2021 The new era of proton therapy 175
Chapter 9 The future of proton therapy at PSI 199
Index 211
About the authors 217

National foreword

By Daniel Zwahlen, Head and Chairman of the Department for Radiation Oncology at the Cantonal Hospital Winterthur, President of the Swiss Society of Radiation Oncology

Growing understanding of the fundamental properties of matter and energy has over the years ignited the development of many of the healthcare technologies we take for granted today. It was the Paul Scherrer Institute's work on translating knowledge of basic science into clinical applications, through collaboration between physicists and other disciplines, which made proton therapy as we know it today possible. Without question, the discovery and evolution of particle therapy in all its aspects has pioneered new treatment concepts and opened new avenues for successfully treating cancer patients.

As nicely explained in this book, the Paul Scherrer Institute pioneered modern proton therapy and has become a world leader. Generations of radiation oncologists were inspired by the brilliant and curious minds working together at PSI, and exposure to proton therapy there propelled their skills in highly specialised cancer treatments. This unique environment will hopefully continue to engage future generations of radiation oncologists in finding better ways of treating cancer through innovative technology.

In science and medicine there is always a need for big and ambitious projects, a place where visionary people come together in the right environment to put theory into practice. Needless to say, the Paul Scherrer Institute has proved itself to be such a unique place in Switzerland.

As a radiation oncologist myself practising in Switzerland, it is a privilege to be part of this community and have access to this unique and evolving technology for cancer patients in need.

In the name of the Swiss Society of Radiation Oncology, I congratulate the current leadership for their initiative and visionary approach to constantly improving cancer treatment. But let us not forget the past generation of great minds and leaders at PSI that made these advancements in curing cancer possible.

International foreword

By Jay Flanz, Project Director and Technical Director of the Burr Proton Therapy Center, Massachusetts General Hospital, Past Chairman of the Particle Therapy Co-Operative Group

The early development of conventional radiation therapy started in only a few institutions. It was initially believed, in the 1930s, that such formidable installations (with super-voltage generators) would be prohibitive for the average radiologist. Therefore it would be limited to those institutions with engineering and physics skills available and should be centralised.

The development of charged particle therapy followed a similar path. At first, it was only possible to offer it in the environment of a national laboratory with the resources to produce a medically useful beam. There it might have stayed had there not been true interest in developing medical potential. Researchers explored the most suitable particle for treatment: there were neutrons at Fermilab (US), pions at Los Alamos (US) and PSI, protons at Harvard (US), PSI and Uppsala (Sweden), and protons and heavier particles at Berkeley (US) and the National Institute of Radiological Sciences in Chiba, Japan.

What enabled growth outside the laboratory was, in the first place, the realisation that protons – the easiest of all the charged particles to produce and control – were a viable therapeutic tool. Technology sharing from laboratory to industry was also important. But the continued development of charged particle therapy techniques and technology by laboratory staff was crucial. And nowhere were the staff more thorough and prolific than at PSI.

While, initially, industry focused on efficacy and sales, treatment-related developments were more likely to happen within the scientific community. More recently, it has been unusual for a national laboratory such as PSI to invest resources in application-specific investigations that are not part of its mainstream research programme. Why, then, did PSI put the effort into developing proton beam scanning, compact gantries, treatment planning, imaging external to the room and a superconducting cyclotron?

The answer lies within the pages of this book, which documents PSI's strategy of considering public benefit and assembling a remarkable staff to achieve that. Other laboratory facilities conducted a clinical programme too, but the support that was provided at PSI was and is unprecedented. This is not to say that PSI did not benefit from interaction with other groups, such as those at Berkeley, Chiba and Los Alamos. It is very hard to develop a field with limited resources, but if an important resource shows and shares the way to what works and what doesn't, it lifts the state of the art of the field years ahead.

The personalities and mentalities of the individuals who spearheaded the path at PSI are unique indeed. This book provides important insights into the people and how they managed their circumstances. PSI staff and management – people like Eros Pedroni, Martin Jermann, Tony Lomax, Gudrun Goitein and Damien Weber – saw and worked towards a future that few others envisioned. It wasn't just a case of physics versus cancer; it was also physics and medicine collaborating against cancer. Other centres in Japan, the United States and Germany also developed particle beam scanning systems, but PSI developed the first scanning system to be put into full-time clinical operation. Subsequently they developed an impressive state-of-the-art scanning system in Gantry 2. At a time when there was only one main type of proton therapy cyclotron, PSI staff played a major role in developing a superconducting cyclotron and opened the door for proton therapy facilities to grow.

This is more than simply interesting. It is vital for posterity to capture the mindset and learn about the people who contributed to so much growth in the field. Who they are and what they did can serve as motivation for young professionals to emulate and seasoned practitioners to admire. For a field to grow, and for the community of patients to be best served, it's necessary to learn, grow, improve and share, the way it has been done at PSI. This book provides a look at that extremely important perspective. I, for one, read it with great pleasure.

Introduction: about this story

Patients coming to PSI may be aware that they are being treated in one of the leading proton therapy centres in the world. But visitors may be less aware of the important part that PSI has played in pioneering proton therapy globally, developing techniques that have made it the well-established and effective cancer treatment that it is today.

This is the place that pioneered a technique of delivering protons to tumours called pencil beam scanning (also known as spot scanning). The technique, routinely used at PSI since 1996, is now utilised in nearly every proton therapy centre in the world. It is, in the words of Damien Weber, the current head of the Center for Proton Therapy at PSI, "a beautiful way to paint radiation onto a tumour and to decrease the dose of radiation delivered outside the target".

This is the place that developed a completely new paradigm for proton therapy treatment called "intensity modulated proton therapy" (IMPT), which enables the correct dose of protons to be more accurately targeted to every part of a tumour – again, now used as standard around the world.

This is also the place that perfected techniques for delivering protons to eye cancers; that realised complex new gantry systems for delivering treatments; that was the first to administer proton pencil beam scanning to adults and very young children.

Such complex and infrastructure-dependent innovations would not have been possible in a hospital. The Paul Scherrer Institute, where the Center for Proton Therapy developed, is a scientific centre of excellence in which the priority is pure science. It revolves around discovery and innovation.

So this book, the story of how proton therapy developed at PSI, is a story of the successful application of pure science. It is also a story about people – the engineers, physicians and physicists who came together at PSI to learn from each other and apply high physics to the practicalities of human biology. From the moment a medical clinician first took up a role at PSI in the 1980s, the quest for medical excellence began.

Gudrun Goitein, who was medical director at PSI's Center for Proton Therapy between 1989 and 2006, and again between 2011 and 2013, sums up the uniquely collaborative and productive atmosphere that has thrived at PSI.

"If one looks at the entire evolution of radiation therapy, one realises that it was almost always driven by the physicists and engineers – who offered something that might be feasible for the physicians to take up. And if the physicians were interested in the physics and understood what they were working with, then it came out well for patients.

"If you are in an environment like PSI, medical disciplines really cannot help but understand the science they are working with. For me that's essential. In surgery, if you have a surgeon who doesn't understand what a knife is, you are lost.

"So if you look at the development of particle therapy, you can see that the coming together of physicists and physicians was the driving force behind progress. For me, this was the characteristic of the whole project at PSI."

This book tells how these driven, ambitious men and women – physicists, engineers, radiation oncologists and other clinicians – worked together over the course of half a century at PSI to push boundaries.

Martin Grossmann, who came to PSI in the early 1990s and was responsible for proton therapy control and safety systems, sums up the secret of PSI's success:

"When people first thought of proton therapy in the 1950s, the technology was not yet ready to fully develop it. But it became available, and at PSI we brought together the people with experience in these technologies, who realised how to apply them. The main message is that innovation is about putting technology together with people.

"This is true for our control systems, our particle accelerator, the beam line, the proton therapy gantries. And the same applies to the medical doctors who were ready to work together with physicists and engineers. This is really why PSI was so successful and led the field for so long."

The course of scientific and medical progress is rarely smooth, and these people continued in the face of failures and outside scepticism. So central to the story of proton therapy at PSI are also the directors and administrators, such as Jean-Pierre Blaser, Meinrad Eberle and Martin Jermann, who had the courage to press forward with innovations in particle therapy despite pressure to stop and spend money on more physics-orientated projects.

"Even when we started clinical treatment in 1996 there was a lot of scepticism, and it took about eight years to convince the community that this was the right way to go," says Martin Jermann, who became a member of the PSI directorate in 1988 and was programme manager for proton therapy between 1999 and 2013. *"But if you look today, there are around 100 particle therapy units in clinical operation, and 80 to 85 per cent are using the pencil beam scanning technology we devised."*

Today, a quarter of a century on from PSI's first revolutionary use of these scanning techniques to treat human cancers, this book celebrates the Center for Proton Therapy's relentlessly inquiring spirit. It is written not for experts or academics, but for anyone with an interest in scientists' constant quest to find out more, overcome more and accomplish more – not only to expand knowledge but to help fellow human beings.

Some important background: radiation therapy

This short outline of radiation therapy provides explanatory background for the chapters that follow.

What is radiation therapy?
When we talk about radiation therapy (also known as radiotherapy), we usually mean "external beam radiation therapy" – focusing radiation on a target cancer to kill or damage abnormal cells. There are also other, less commonly used, types of radiation therapy involving implants, injections, capsules, drinks and surgical techniques.

Simply put, radiation is the transmission of energy carried by particles from place to place. In the context of external beam radiation therapy, it means using beams of these particles to deposit high energies into a tumour. Radiation oncologists use these beams to damage the DNA in cancer cells, causing them to die or become unable to reproduce.

To effectively treat cancer, the particles in the beam need to have sufficient energy to cause damaging ionisation within the tumour's cells. Ionising radiation can come in the form of neutral particles, such as photons and neutrons, or accelerated charged particles, such as protons, electrons and heavy ions.

For more than 100 years, scientists and physicians have explored the use of different types of particle to deliver radiation therapy effectively – and as safely as possible – to the patient.

How radiation therapy works
Radiation therapy is a "local" therapy – meaning that, unlike chemotherapy or drugs, it acts in one location rather than throughout the body's systems. Its great advantage is that the radiation dose is predictable, measurable and reproducible. To kill or neutralise a cancer, the radiation should, as far as possible, cover the entire tumour with a precisely prescribed dose.

Radiation therapy can be highly effective at destroying cancers because fast-reproducing cancer cells are more susceptible than normal cells to the damaging effects of ionising radiation. The goal is always the same: to administer a sufficiently high dose to the tumour, but as little as possible to the surrounding healthy tissue.

This is not always simple. Some cancers need particularly high doses of radiation to neutralise them – and the chances of severe side effects are great, sometimes so great that the risks of radiation treatment outweigh the benefits.

So the quest for better types of radiation therapy has centred on addressing this delicate balance: maximising damage to the tumour, minimising damage to the rest of the body. The holy grail is a beam powerful enough to kill or slow down all cancers, but sufficiently targeted and contained to prevent radiation damage to all neighbouring tissue.

The right particle for radiation therapy
Whether cells survive radiation depends primarily on the amount of radiation given (the dose is usually measured in a unit called a "gray", abbreviated to Gy). The kind of radiation and the type of particle forming the beam also play essential roles.

Different types of primary particle can have different characteristics in terms of how well they penetrate the body, spread, produce secondary particles and finally ionise cells and tissue. These different characteristics also affect how the beams can be shaped to irradiate the full three-dimensional "volume" of a tumour.

Beams of particles can have complex effects that produce different "qualities" of radiation. A commonly used distinction is between "high LET" and "low LET" radiation, where LET stands for "linear energy transfer" (see "What is linear energy transfer?" box). This aspect is particularly relevant for PSI's pion therapy project (see Chapters 1 and 2).

What is linear energy transfer (LET)?
If a charged particle goes through matter at high speed, there are many interactions along its path. As a result, the particle is slowed down and continuously loses energy. This energy loss with distance is called linear energy transfer (LET) and depends on three things: the medium being passed through, the type of particle and the velocity of the particle. LET increases with decreasing velocity. This is a very important consideration when targeting radiation at tumours (see "About Bragg peaks" box).

> The amount of LET also affects the biological effect on the tissue along the particle track. A high LET particle produces a higher density of interactions along the track, and for living cells this leads to significant damage. A particle with low LET, on the other hand, is likely to cause only limited damage, meaning that living cells are more likely to continue functioning or be repaired.

Today's "conventional" radiation therapy uses accelerated electrons to hit a metal target, which produces photons (also known as X-rays). A beam of these photons can be shaped to the form of the tumour being targeted. The machines that produce the photon beams are called medical LINACs (linear accelerators), and they were introduced in the 1950s.

This kind of radiation therapy using photons is effective, widely available and comparatively economical. Photons are low LET particles. However, when a cancer is deep within the body, it is not always easy to conform a photon radiation dose to the shape of the tumour – a situation that is normally addressed by using combined beams from multiple directions.

Other subatomic particles – such as protons, neutrons, electrons, ions and pions – have different combinations of characteristics in terms of their energy and their ability to be targeted. All have been explored to see whether they have advantages over photons.

Radiation therapy with protons
The potential of accelerated protons as particles for radiation therapy was first recognised in the 1940s by scientists at Harvard University in the USA, who concluded that the proton had particular qualities that might make it a good means of conveying high energy to tumours. Researchers had already experimented with accelerated electrons as high energy particles for radiation therapy. However, being light, electrons tended to scatter when passing through tissue before reaching their target – which was a problem if the target was deep-seated in the body.

Protons, on the other hand, being heavier particles, scattered less and went through tissue in straighter lines, so were more easily targeted. When passing through tissue, they yielded their maximum energy immediately before stopping. This meant that the maximum dose could be released at the target, leaving the tissue that the beams passed through subject to much smaller energies, and therefore less likely to be damaged (see "About Bragg peaks" box).

About Bragg peaks

Different types of particle passing through tissue yield their ionising radiation at different points in their journey. The point at which they release most ionising energy is called the "Bragg peak". In radiation therapy, the ideal is for most radiation to be released when the particles reach the tumour. The graph shows how photons (used in conventional radiation therapy) produce a high dose at first, which gradually diminishes as they pass through tissue. In contrast, protons (in this case administered through the "spot scanning" technique developed at PSI) release their maximum dose much deeper within tissue – mostly at a specific point as they slow down and stop (the Bragg peak). The position of the Bragg peak can be controlled by changing the beam energy at the point of entry.

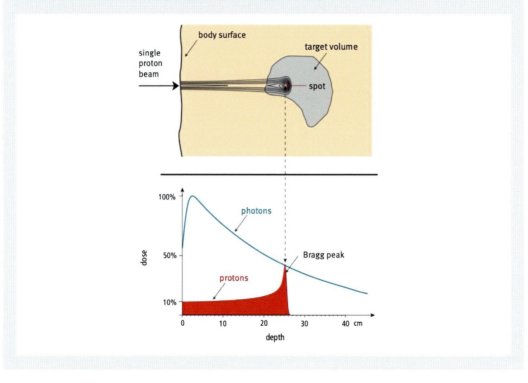

The first human patients were treated with proton beams in the 1950s, at the Berkeley Laboratory, California, and the Gustaf Werner Institute for Nuclear Chemistry in Uppsala, Sweden. In the late 1970s, when particle accelerators and computerised delivery technology improved, there was a new wave of interest and investment in using protons for cancer treatment.

Proton therapy, using beam delivery techniques developed at the Paul Scherrer Institute, is now practised at over 100 treatment centres around the world. Because it hits the tumour very precisely with high doses and protects the surrounding tissue as much as possible, proton therapy is especially suitable for tumours located near sensitive organs, for example in the head or pelvis or on the spinal column. Higher doses of radiation can be delivered to the tumour, so that even previously radiation-resistant cancers can be killed or neutralised.

Proton treatment is particularly useful in the treatment of children. Because children are highly sensitive to radiation, conventional, less targeted, radiation therapy can bring major side effects. The accuracy of proton therapy, in contrast, means that children usually have fewer radiation-induced long-term complications.

Around 250,000 people in the world have been treated with proton therapy since the 1980s.[1] They are often referred for proton therapy when treatment of their tumour is challenging – because it is resistant to conventional radiation therapy, or because other techniques would prove too damaging to surrounding tissue.

The procedure of radiation therapy
People who need radiation therapy for a tumour are referred to a radiation oncologist, who assesses all the diagnostic information about the patient. The radiation oncologist will require detailed images – obtained through X-rays/CT, PET or MRI scanning – which will allow them to delineate the volume to be treated and define the regions at risk that must be spared from radiation.

The treatment planning process also requires a medical physicist or dosimetrist to optimise the settings on the treatment device (such as a LINAC) in order to best match the radiation oncologist's prescription. The process involves a "simulation", during which the patient is placed in a treatment position, sometimes with immobilisation devices such as masks, headrests and body moulds.

The result of the preparation is a therapy plan containing the technical information for the treatment device: how to move the beam and the patient in order to fully cover the target volume with the intended dose. Control systems ensure the treatment device strictly follows the therapy plan.

---

[1] Patient statistics, Particle Therapy Co-Operative Group (PTCOG), https://www.ptcog.ch/index.php/ptcog-patient-statistics.

Radiation therapy is usually delivered to patients in "fractions" – in other words, in several separate sessions in which a proportion of the overall dose is delivered. This is extremely important because although healthy cells recover from radiation better than cancerous cells, they still need time to recover. After several fractions, the net effect is that the healthy cells have recovered, but the cancerous cells have not. This would not be the case if the full dose was given in one go.

About proton therapy at PSI
The Paul Scherrer Institute (PSI) is the largest research institute in Switzerland for natural and engineering sciences, conducting cutting-edge research in three main fields: matter and materials; energy and the environment; and human health. Every year, more than 2,500 scientists from Switzerland and around the world come to PSI to use its large research facilities to carry out experiments. PSI employs about 2,100 people altogether.

Within PSI is the Center for Proton Therapy, which employs 120 people and is one of only 34 proton therapy centres in Europe (as at 2020). Patients who are judged to be eligible for proton therapy are referred there by the doctors from the hospitals treating them.

Since 1996, the Center for Proton Therapy has treated over 2,000 patients with deep-seated tumours. It is also a world leader in treating tumours of the eye with protons. No other institution in the world has treated so many ocular tumours – around 7,800. This irradiation takes only four days and in 90 per cent of cases the patient's eye can be saved. A recent survey found that approximately 21 per cent of all eye melanoma patients treated with protons worldwide have been managed at PSI.

A timeline of particle therapy at the Paul Scherrer Institute

1954	First therapeutic exposure of humans to protons, Berkeley Laboratory, USA
1968	Swiss Institute for Nuclear Research (SIN) founded (later to become Paul Scherrer Institute)
1973	First proton beam extracted from main accelerator at SIN
1974	First radiobiological experiments on pion beams begin at SIN
1975	First ocular tumours treated with proton beams in Boston, USA
	Construction of the SIN piotron and biomedical facility begins
1979	First patients treated with pions in Vancouver (TRIUMF)
1981	First patients treated with pions at SIN
1984	First proton treatment of patients with ocular tumours treated at SIN's OPTIS facility
1988	Paul Scherrer Institute founded by merging SIN and Institute for Reactor Research
	First experiments in dynamic spot scanning conducted at PSI
1989	Proposals for Gantry 1 spot scanning project first put forward
1990	PSI Director Jean-Pierre Blaser retires
1991	1,000th patient with an eye tumour treated at the OPTIS facility
1992	Horizontal beam line dedicated to spot scanning comes into operation
	Meinrad Eberle becomes PSI Director
1993	Piotron decommissioned, having treated 503 patients
1994	First beam through Gantry 1
	Animal proton therapy programme begins
1996	First human treatment using spot scanning in the world on Gantry 1
1998	Evaluation of proton therapy strategy at PSI
	Martin Jermann mandated to lead further development of proton therapy at PSI
1999	PSI directorate decides to commit to an expansion of proton therapy
	PROSCAN project starts
	First patient in the world treated using intensity modulated proton therapy (IMPT) at PSI
2000	PROSCAN project formally launched

2001	Swiss government approves health insurance coverage for proton therapy for specific tumours
	Contract signed for ACCEL to construct superconducting cyclotron (COMET) to agreed specifications
2002	Gantry 2 first conceived
2003	Preparation of infrastructure for new proton therapy facility
2004	Paediatric programme begins, and first young children treated with proton therapy under anaesthesia on Gantry 1
2005	First beam extracted from new COMET cyclotron
2007	Proton therapy becomes a year-round operation at Gantry 1
2008	Swiss Canton of Aargau donates 20 million Swiss francs for proton therapy at PSI
	First beam sent through Gantry 2
2010	Center for Proton Therapy extended
	OPTIS2 facility treats first patient
2012	Swiss Canton of Zürich donates 20 million Swiss francs for Gantry 3
	Gantry 3 project begins
2013	First patient treated on Gantry 2
2014	30th anniversary of OPTIS: 6,300 eye patients treated during that time
2015	1000th patient treated on a PSI gantry
2016	First paediatric patient treated on Gantry 2, under anaesthesia
2018	Patient treatment on Gantry 3 begins, patient treatment on Gantry 1 ends
2021	390 children treated with proton therapy under anaesthesia since 2004
	FLASH experiments start at Gantry 1
2021	First lung cancer patient treated with gated/multiple scanned proton therapy
2022	Forecast of 10,000 patients in total treated at PSI

Chapter 1

1968–1977
Beginnings: the piotron

A patient approaching the Center for Proton Therapy at the Paul Scherrer Institute (PSI) today might be forgiven for feeling a little intimidated. It is a friendly and comfortable place once you have walked through the doors, but first impressions are nothing like the hospital-based treatment facilities most people are likely to have experienced.

PSI sits in solitary splendour on both banks of the river Aare in the Swiss municipality of Villigen, surrounded by forest, fields and hills. The Center for Proton Therapy is located within the array of industrial-style buildings that form Switzerland's leading science research centre – among them the massive circular "doughnut" housing the Swiss Light Source, which provides photon beams for research in

Figure 1. An aerial shot of the Paul Scherrer Institute. The circular construction is the Swiss Light Source. To its left, the large light-grey block is PSI's experimental hall, which incorporates the Center for Proton Therapy's gantries and cyclotron. The attached low-level medical building is visible to its right.
Picture: Paul Scherrer Institute

materials science, biology and chemistry. The nearest town is Brugg, 9 km away. The setting is at once beautiful and formidable.

The environment of the Center for Proton Therapy is a mark of its unique origins, at the heart of an institute set on uncovering the nature of matter and finding ways to apply that understanding for the benefit of mankind. The subatomic particles produced by the accelerators at the institute were first researched as treatments there in the 1970s, and the treatment centre has remained within the context of big science ever since.

The origins of the Paul Scherrer Institute
The Paul Scherrer Institute started out as the Swiss Institute for Nuclear Research (SIN), founded in 1968 as a national centre dedicated to basic physics research. Building began in 1969 in Villigen, across the river Aare from the Swiss Federal Institute for Reactor Research (EIR).

SIN was established at a time when governments around the world were investing in facilities to increase understanding of the subatomic particles that it was now possible to produce using particle accelerators. Since the 1950s, increasingly powerful accelerators had made manifest more and more elementary particles previously only conjectured about, among them the meson and pi-meson (or pion).

The next step was to understand these particles better and investigate whether they might have useful applications. So SIN was established as Switzerland's "meson factory", at around the same time as similar facilities were being built in America (the Los Alamos Meson Physics Facility, LAMPF, in California) and Canada (the Tri-University Meson Facility, TRIUMF, in Vancouver). Both of these research centres were to go on to build medical facilities linked to their particle accelerators in the early 1970s, and like SIN explored the treatment applications of the subatomic particles called pi-mesons, or pions. Physics research centres in Japan also joined the quest for better means of particle therapy in 1977.

Investigating the potential medical use of pions was always part of the plan for SIN. In the words of founding director Jean-Pierre Blaser, one of its core functions was to produce mesons and other subatomic particles "in quantities enabling us to apply them . . . as tools in nuclear and particle physics as well as to open a range of new applications like muon spin rotation and pion cancer therapy".

Blaser, who went on to become director of PSI on its formation in 1988, had been a pupil of Paul Scherrer, the influential Swiss physicist after whom the institute was later named. He saw SIN as a national resource for pure physics, where universities and other research groups and institutions could freely investigate science for its own sake. But he also believed that the new breakthroughs in particle physics should be explored with human applications in mind.

About subatomic particles
Atoms are composed of many self-contained units called subatomic particles. They include electrons, which are light and negatively charged, as well as the heavier building blocks of the atom's dense nucleus – positively charged protons and electrically neutral neutrons.

But these are not the only subatomic particles. Particles called mesons have a mass between that of the electron and the proton. There are many types of meson, including the pi-meson, which carries the force between protons and neutrons. Protons, neutrons and pions themselves are made up of elementary particles called quarks and gluons, and the electron is only one member of a class of elementary particles that includes the muon and the neutrino. The study of these particles was at the heart of work at SIN, and is still carried out at PSI.

About pions
Pi-mesons (abbreviated to "pions") are subatomic particles produced in the high-energy collisions of accelerated protons, and they are considerably heavier than electrons. There are three types of pion: those with a positive charge, those with a negative charge and those without a charge. It was the properties of the negative pion that fascinated medical science.

Special qualities of protons and pions
Since the early 1950s, researchers in the United States and Sweden had been exploring the use of accelerated protons to treat cancer. Being heavy particles, protons went through tissue in straight lines (unlike electrons, which were easily scattered), so were more easily focused onto a tumour. When passing through tissue, they yielded most of their energy immediately before stopping, meaning that the maximum dose was released at the target, and the skin and tissue that the beams passed through to reach the target were less likely to be damaged (see "About Bragg peaks" box, page 18).

While the characteristics of the proton made its potential for radiation therapy promising, the new ability to create pi-mesons, or pions, raised another option. Pions are produced when an accelerated proton hits matter. Like protons, pions are heavy – 208 times heavier than an electron. So it was clear that they potentially had targeting advantages similar to those of protons, concentrating their energy dose where it was needed. The therapeutic potential of pions had actually been predicted as early as the 1950s by Nobel Prize-winning physicist Enrico Fermi.[2]

But in addition, pions had a secondary advantage. Negative pions seemed to release extra explosions of energy when they arrived at their target. Eros Pedroni, the physicist who was a driving force on the pioneering pion projects at SIN, explains:

"The exciting idea raised in the 1960s to use negative pions for radiation therapy was based on very peculiar properties of these particles when they deposit radiation in matter. The dose of the negative pions is essentially delivered in two consecutive physical processes.

"When a negative pion passes through a material, it first loses energy by collisions with the electrons in the material, and slows down until it stops at a well-defined position. The range of a pion beam in materials is governed by the initial energy of the beam. It shows a behaviour typical for all charged particle beams, with an increased dose deposition towards the end of the trajectory producing the so-called Bragg peak.

"This localisation of the dose in depth is more favourable with charged particles such as protons, pions and ions than with neutral particles, such as neutrons or the photons used in conventional X-ray therapy.

"The second process, and this is very distinctive to this particle, is the formation of a so-called 'pion star'. When the negative pion reaches a standstill, it is captured by the positive charge of the nucleus of an atom of the material it has stopped in. This creates a pionic atom. The heavy negative pion collides immediately with the nucleus, and the energy stored in the mass of the pion is released in the nuclear collision. The nucleus explodes, with the emission of nuclear fragments (the 'star'

---

[2] As told to Carl von Essen, see Chapter 2.

formation). Most of them are short-ranged, densely ionising, high energy particles. The dose deposited by these fragments is the so-called 'star dose'.

"Jean-Pierre Blaser introduced to us a pictorial analogy for a targeted pion: it was like a shrapnel bullet with a retarded explosion, investing its explosive only where it mattered."

The star dose is not simply a magnification of the Bragg peak. The quality of its radiation is very distinctive: it contains high LET radiation. High LET (linear energy transfer) radiation is known to have a very high cell-killing efficiency. It wipes out the usual capacity of cells to repair radiation damage. So the focus of trials examining the use of neutrons and pions was the eradication of inherently "radioresistant" tumours.

Pions presented an exciting prospect because they could deliver the high LET radiation only where it was needed: in the tumour. The dose delivered before the pions reached their target was low LET radiation, and healthy tissue in the vicinity could be efficiently protected by delivering in fractions – as with conventional therapy. What was not known, however, was what happens to non-cancerous tissue within the target area.

This confinement of the high LET radiation to the target gave pions the advantage over neutrons when it came to clinical potential, says Eros Pedroni: *"The neutron dose falls off exponentially as it passes through tissue. Neutrons deliver a much higher dose burden to healthy tissues than pions, and the high LET is equally damaging within and outside the target. With neutrons, there is a 'no repair' situation everywhere. The clinical use of heavier ions like carbon, which are considered today as potentially the best approach for applying high LET radiation therapy, were in the 1970s still decades away."*

About particle accelerators
Particle accelerators propel subatomic particles which have a charge – such as protons, electrons and ions – at speeds almost up to the speed of light. They normally do this by means of electric fields arranged so that they accelerate the charged particles through a vacuum tube, building their speed and forming them into a beam as they go. There are different types of accelerator, each producing a particle beam with specific properties.

Particle accelerators were originally developed for basic scientific research – researchers smash accelerated subatomic particles against targets to release smaller subatomic particles. But they now also play an important role in health, industry and security.

The linear accelerator (LINAC) and the circular accelerator are the two main types. In linear accelerators, often used to accelerate electrons, the electric fields are arranged along a straight path. Circular accelerators, particularly a type called a cyclotron, are often used to generate proton beams for medical purposes.

In cyclotrons, protons are kept on a spiral path by a strong magnetic field. This allows the protons to be accelerated by the same electric field several times, gaining energy with each round, until they reach the outermost point and are extracted. Cyclotrons therefore have a compact size.

Protons for the cyclotron are sourced from hydrogen gas. An electric field is used to strip electrons from hydrogen atoms, leaving single protons, which are then fed into the cyclotron.

First pion experiments at SIN
The accelerator being built at SIN was a 590 MeV cyclotron capable of accelerating protons to sufficient speeds to produce pi-mesons. It had an ingenious ring-shaped design, so was named the "ring cyclotron". Another smaller injector cyclotron was used to get the protons up to speed before feeding them into the main ring cyclotron.

By 1970, the meson factory in Los Alamos was already planning an experimental therapeutic unit with a view to treating patients with beams of pions for the first time. There was an exciting opportunity to bring Europe into the game. Jean-Pierre Blaser's interest was bolstered by the enthusiasm of the Swiss physicist Charles Perret, who was designing and directing the construction of SIN's main experimental hall, with particular responsibility for radioactivity shielding and radiation protection protocols. Perret had been inspired when he visited the treatment facility linked to the cyclotron at Harvard University, which had begun treating pituitary cancer patients with protons in the early 1960s. Why was a similar Swiss project using particles for treatment not possible, he wondered, based around SIN's new accelerator?

Figure 2. SIN staff gather around the new ring accelerator in 1973. Picture: Paul Scherrer Institute

So, as the SIN facility took shape in the early 1970s, a beam line specifically dedicated to biomedical research was constructed. The first experiments on the effects of a pion beam on tissue were conducted in 1974 by Hans Blattmann and Hedi Fritz-Niggli – experts in radiobiology and dosimetry from the University of Zürich's Radiobiological Institute – at around the same time as researchers in Los Alamos were beginning pilot studies using pions on patients.

"We were invited by Jean-Pierre Blaser to come and work at SIN," says Hans Blattmann. *"I started working with pion beams, and in the beginning we worked with cells only, because it was necessary to understand the dose distribution in this vertical beam."*

However, the SIN cell-culture studies gave rise to concerns that a single fixed pion beam might not provide enough beam intensity to successfully treat deep-seated tumours. International workshops held at SIN in 1975 considered the problem. Informed by these, Georg Vecsey, an expert in superconducting technology at SIN, came up with an answer. They should build a "superconducting double torus spectrometer", based on a prototype device built at Stanford University. It was a kind of particle accelerator which, by means of enormous magnets, could direct pions towards the patient and deliver them in a circular arrangement of beams. It would deliver pions more efficiently than a single fixed beam, and the radial beams would give flexibility in dose delivery.

The idea was received enthusiastically by Blaser, but the project would need more space. Blaser gave the go-ahead for the building of a separate biomedical facility, outside the main SIN experimental hall. Plans for the new pion applicator began to be drawn up.

About proton accelerators at SIN and PSI
Sometimes two or more accelerators are used in series to accelerate protons to high energies with particularly high beam intensity. This was the case for the high-intensity proton accelerator that was constructed as the core facility at SIN in the 1970s. The source of protons, and the first acceleration stage, was a small Cockcroft–Walton accelerator. This fed protons into a small ring cyclotron which acted as an "injector" to the main ring cyclotron. This injector cyclotron (known as Injector 1) brought the protons up to 37 per cent of the speed of light, and to an energy of 72 MeV (mega electron volts, a measure of kinetic energy) before injecting them into SIN/PSI's large ring cyclotron. With a diameter of 15 metres, this large cyclotron accelerates protons to almost 80 per cent of the speed of light and to an energy of 590 MeV. With a beam current of up to 2.4 mA and a resulting power of 1.4 MW, it remains one of the most powerful proton accelerators in the world.

Until the first decade of the 21st century, this large ring cyclotron was the source of protons for proton therapy to treat tumours deep-seated in the body, while the Injector 1 cyclotron was the source of protons for proton therapy for eyes. The proton therapy programme had to share the beam with other users. Today, the proton therapy programme has its own dedicated accelerator – the Compact Medical Therapy cyclotron, or COMET. It is based on a superconducting magnet, which allows the generation of a much stronger magnetic field. Protons are accelerated in a smaller circuit, which results in a more compact machine with a diameter of just over three metres.

Vecsey's plans
Georg Vecsey was a young engineer, originally from Hungary – one of the dozens of émigrés from communist countries in the 1960s and 1970s who found a welcome

home for their scientific skills at SIN. He quickly became the leading specialist for cryogenics and superconductivity at SIN and played a central part in the construction of the ring cyclotron, developing a superconductive magnet for channelling muons. Initially, sceptics had believed his muon channel design was too risky, but it proved highly successful and was eventually adopted by all other meson factories in the world.

"He understood the enormous potential of this innovative magnet technology and developed especially original concepts," wrote Jean-Pierre Blaser in a history of SIN.[3] Now he was designing a pion generator. Vecsey visited the physics department at Stanford to gain a full understanding of the concept.

But this was still an innovative and risky departure. At its simplest, the superconducting spectrometer was a machine for producing pions from an accelerated proton beam and concentrating 60 radially arranged pion beams on a tumour. It had to deliver a sufficient intensity of pion beam to be able to treat tumours within a reasonable fraction time. This was only feasible through the use of superconducting magnets to guide the pions.

Protons accelerated by SIN's ring cyclotron were to be injected into one side of the spectrometer, where they would hit a gas-cooled metallic "production target", resulting in the production of pions. The pions were to be collected, fed into streams, transmitted and focused onto the patient at the other side, five metres away, via two huge superconducting magnetic coils. Between the injecting proton beam and the patient, a massive steel shield would provide protection from unwanted radiation.

Using such large superconducting magnets to collect the pions and direct them at acute angles around the machine was a new and challenging technology. Each toroidal set of 60 coils weighed 6–8 tons and had to be mounted within a large cylindrical vacuum tank. The coils were cooled with liquid helium at a temperature of minus 270 degrees Celsius.

Although SIN was able to finance the building of the machine from its research budget, it could not cover the construction cost of the three-storey biomedical

---

[3] Andreas Pritzker, *The Swiss Institute for Nuclear Research SIN* (Norderstedt: Books on Demand, 2014).

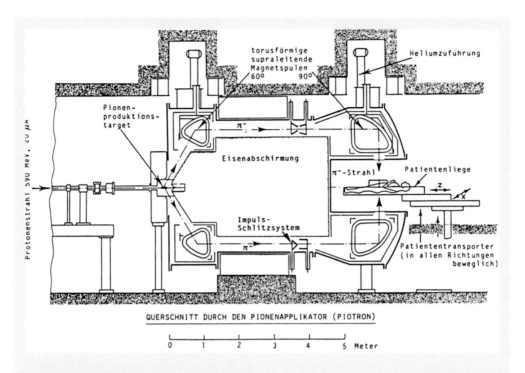

Figure 3. The original design for the pion-generating superconducting spectrometer. Key to the German labels: *Protonenstrahl* 590 MeV 20 µA = proton beam 590 MeV 20 µA; *Pionenproduktionstarget* = Pion production target; *Torusförmige supraleitende Magnetspulen* = Toroidal superconducting coils; *Eisenabschirmung* = Iron shielding; *Impuls-Schlitzsystem* = Momentum slit system; *Heliumzuführung* = Helium supply; *π–Strahl* = pion beam; *Patientenliege* = Patient couch; *Patiententransporter (in allen Richtungen beweglich)* = Patient transporter (movable in all directions).

facility it needed around it. This was to include biology laboratories and rooms for patient examination, treatment, planning, controls, computers and dosimetry.

The Swiss Cancer League came to the rescue and endowed two million Swiss francs. The Swiss government also provided support, and the Federal Office of Public Health gave legal approval for the development of clinical operations at SIN. Excitement about the potential of such a radical medical departure was spreading.

Construction of the superconducting spectrometer and the new biomedical facility began in 1975. In the following year the centre was served with its own dedicated proton beam line – which would prevent SIN's medical work having to compete with other physics experiments. But construction of the beam line was by no means straightforward. To feed the spectrometer with protons, the main 590 MeV beam from the ring cyclotron had to be divided into two using an electrostatic high-voltage "septum", or splitter. Given the high intensity of the beam, this was a significant achievement requiring innovative technology. It would prove to be a very important component for SIN and then PSI, enabling later scientific developments such as the construction of the first proton therapy "gantry".

As the pi-meson programme took shape, so it attracted some of the key skills and brilliant minds that would set the course for the institute's groundbreaking developments in particle therapy over the next 30 years.

Charles Perret jointly led the project with Georg Vecsey. Working alongside them was Hans Blattmann, who had left the University of Zürich to become the leader of the physics group at SIN and an enthusiastic advocate for the pion therapy project. But there was another vital part of the equation to fill in. If SIN was to investigate the potential of particles for treatment, it needed more than physicists and engineers: it needed a physician.

A radiation oncologist joins
Moves to employ someone to guide and take responsibility for the clinical side of the programme started in 1974, as the first experiments in the biomedical beam area began. Carl von Essen, chairman of the Radiation Oncology Department at the medical school at the University of California San Diego (UCSD), saw the job advertisement and decided to apply. At the time, he was on sabbatical at the Swiss Institute for Experimental Cancer Research (ISREC) in Lausanne.

As early as the 1950s, when he worked on electron therapy research at Stanford, von Essen had been interested in the potential of new particles to break the bonds of conventional radiotherapy. His interest in pions had been fired in 1950, when he met the physicist Enrico Fermi, creator of the world's first nuclear reactor and known as the architect of the nuclear age.

"He thought that unique qualities of the pion might be of advantage in treating cancer," says von Essen. *"He predicted that a useful beam of pions could be generated some 10 or 20 years later, and so it was."*

With his interest in pions aroused, von Essen had acted as consultant to the Los Alamos Meson Production Facility, designing its clinical studies of pions. Now, the prospect of investigating them further intrigued him.

He was interviewed by Vecsey, Perret, Blaser, Fritz-Niggli and Fritz Heinzel, head of radiotherapy at Triemli Hospital in Zürich. Perret offered the job to von Essen personally in the United States, while the two of them were attending a meeting on particle therapy at the Greek Theatre at the University of California, Berkeley – a short walk from the cyclotron of the Lawrence Berkeley National Laboratory. The contract specified that von Essen should learn German.

It was agreed that von Essen should work part-time at first, flying out to Switzerland for three weeks every other month to plan and develop the pion programme. For the remainder of his time, von Essen would work as a research professor at the University of New Mexico, gaining valuable experience on the university's pion clinical programme at Los Alamos, working under its director, Professor Morton Kligerman.

By 1976 von Essen was working full-time at SIN, where he shared an office with Vecsey and Perret while the new spectrometer and its building were completed. The atmosphere was not always composed.

"I quickly found my colleagues to have strong personalities and divergent views on many issues of the project," says von Essen. *"I found myself acting as a mediator and eventually, as the programme came to fruition, was asked by Jean-Pierre Blaser to take charge. So I became director of the programme.*

"To resolve the skirmishes I assigned Perret to be in charge of all matters involving the design of the treatment couch, patient immobilisation and orientation of the volume to be treated. Vecsey was in complete charge of the spectrometer design and operation."

Carl von Essen's diplomatic skills were needed to build relationships beyond SIN too. In setting up a clinical research programme, it was important to have good networks of radiation therapists, surgeons and medical oncologists in Switzerland – if nothing else, to ensure that patients were referred to the SIN facility. He established an international advisory committee to encourage collaboration, composed of some of the best-known physicists and clinicians in Europe, all prepared to visit SIN for two days, three times a year – all expenses and an honorarium paid.

Carl von Essen: *"We had around a dozen doctors from Edinburgh, London, Cardiff, Rotterdam, Brescia, Paris, Dijon. Jean-Pierre Blaser was the one who encouraged*

Figure 4. Carl von Essen beside the piotron.

all this. It must have been expensive, but I think there was an element of public relations about it."

At each meeting, von Essen and Hans Blattmann would present topics relevant to the clinical programme and invite discussion.

"I relied on others' opinions and my role was often to moderate discussions," he says. *"There was vast interest by Swiss and other European physicists and laity in the potential of this new modality to cure cancer, but we found the Swiss medical establishment rather reluctant to collaborate, and we had to work hard to bring radiotherapists on board."*

It was Carl von Essen who suggested a new name for the superconducting spectrometer: it was to be called the piotron.

The piotron team takes shape

The team engaged to transform the 60 beam lines into a medical therapy system originally consisted of Charles Perret and physicist John Crawford. It was clear to them that sophisticated computerised systems would play an important part in treatment planning. They needed a physicist on the team well acquainted with computer programming.

In 1977, they found Eros Pedroni. A Swiss-Italian from Ticino, Pedroni was one of a group of PhD students who had transferred to SIN from the Swiss Federal Institute of Technology (ETH Zürich) in 1972 to help start up the first physics experiments with the new cyclotron. At the end of his thesis work, he was analysing multi-wire-proportional-chambers data on SIN's central PDP-11/45 computer – at a time when affordable personal computers were still a distant dream (just a year after the first Apple computer had been produced). And it was then that he was approached by Perret.

"Charles Perret asked me if I would be interested in joining this project," says Pedroni. *"It was probably my experience in computers that led him to ask me. But when I saw this piotron and the superconducting machinery, it was very exciting. I found, and I still find, pi-mesons fascinating – and here was a machine that could channel their potential. And immediately came all the questions about the problems of providing a workable beam-delivery technique and the challenging task of developing treatment planning for this unique system of fixed converging beams.*

"Only the so-called pi-meson factories at LAMPF, TRIUMF and SIN were capable of producing pion beams with sufficient dose rate for radiation therapy. Pion therapy was a new research field open for big surprises, especially in view of the peculiar radiobiology of the pion.

"My immediate impression of pion therapy was that it was really something very special. The relevant energy is transported into the tumour by the particle mass. Was this not another way of showing the revolutionary power of Einstein's formula $E=mc^2$?

"I found it a bit intimidating coming onto the piotron project because the people working in this field had such a lot of talent, and as a young man I feared I was not at the same level. But these were really exciting times and I think it was much easier to make progress then. You had responsibility, and if something went wrong it

would come back to you, but you could decide your own direction and go ahead. This was a time for big steps. There was a lot of spirit."

So Pedroni joined as a postdoc in 1977 and began work on the piotron's treatment planning systems. Hans Blattmann and physicist Miriam Salzmann also joined the medical systems team. Their work was to be informed by von Essen.

"Carl von Essen brought a lot of dedication and engagement," says Pedroni. *"He successfully guided the project in the critical first five years, during the realisation and commissioning of a very challenging and highly innovative project. I learned from him the essentials of radiotherapy, and profited from it for the rest of my career."*

That learning experience was to stand the institute in good stead – not just through the years of pion therapy, but through the development of new ways of delivering proton therapy in the decades to come.

Chapter 2

1977–1983
Treating in three dimensions

Carl von Essen says the years leading up to the first piotron patient treatment in 1981 were "probably the most creative of my life".

"It was a heady atmosphere. Everyone had great ideas, sometimes way beyond what was practical. There were many discussions and groups – a really creative time. Everybody was enthused, everybody was having ideas."

But ideas brought the need for decisions, and one of the most important early choices concerned how the pion beams should be delivered to patients. As it turned out, the decision that was made proved crucial: it set the path for the development of a new technique called spot scanning (later called pencil beam scanning), which was to be at the core of PSI's innovation in proton therapy.

The 60 beams in Georg Vecsey's design had to be controlled and coordinated to deliver the most effective dose for each individual patient. Conventionally, radiation therapy was delivered through a single beam of particles, shaped to the target tumour using "collimators" – metal devices, often specially made for each patient. But this was not possible given the innovative design of the piotron, which utilised many identical beams with fixed beam optics, all with the same energy. A new, more dynamic approach had to be found.

As the piotron building took shape in 1977, the group came up with two options. Vecsey's original idea, based on the technique envisaged for the Stanford pion generator, was known as "ring scanning".

For both approaches, preparatory CT (computed tomography) scans would first accurately identify the shape and location of the tumour, and this would form the basis of treatment planning – the same procedure as currently performed for conventional radiation therapy. However, back in the 1970s CT planning for radiation therapy was unheard of in Switzerland, and its use at SIN was in itself a quantum leap.

In ring scanning, the patient would be treated while lying on a cylindrical couch aligned with the axis of all the pion beams. The patient would be moved into the piotron with the centre of the body at the centre of the beams. Vecsey's plan was that the patient would be treated in separate "dose slices" as they were fed into the machine.

For each slice, the range of the pion beams was adjusted to the target by changing the current in the superconducting coils, and those beams whose range would stop outside the target were switched off. In effect, as each "slice" was treated, the Bragg peak would traverse the target in ring sections of changing diameter. The result of all the two-dimensional slices put together would be a three-dimensional dose shaped to the target volume.

Charles Perret was the first to point out a major flaw in this approach. He was worried that contamination of the pion beams with electrons and muons would result in a steadily growing low LET dose where the beams met. The concern was that the system would create large variations of the star dose across the target, which would cause significant difficulties when it came to performing dose-escalation studies and interpreting the results of future clinical trials.

Eros Pedroni proposed a new approach called "spot scanning", which could provide a much better homogeneity of dose quality within the target. The pions from all 60 beams would stop and release their intense LET energy at a single "hot spot" within the tumour. Then the patient would be moved so that the next "spot" could be treated.

For this approach, it was important that each of the pion beams should pass through exactly the same amount of material, so that their Bragg peaks all coincided at the target. To ensure this, the patient would be put into a cylinder and surrounded with packs of material (known as boluses) which formed the equivalent of human tissue. This neat cylinder of tissue and tissue equivalent would sit within a tubular bag filled with water (a water bolus, another tissue equivalent) within the piotron. When the patient cylinder was moved horizontally and vertically within the piotron so that the pion hot spot hit different points in the tumour, the flexible water bolus would flex and allow this – at the same time ensuring that each pion beam was always passing through tissue equivalent and not air.

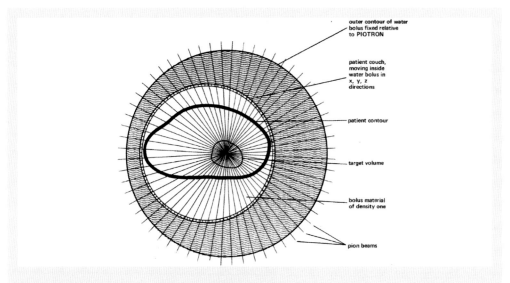

Figure 5. Explanatory diagram of spot scanning within the piotron, published in an article in the International *Journal of Radiation Oncology, Biology, Physics* in 1982. Reproduced with permission.

"The idea is best understood by thinking of treating the patient as if immersed in a horizontal cylindrical water bath," says Eros Pedroni. *"The energy of the 60 beams is chosen for a pion range equal to the radius of the bath. The pion peak of the beams is placed at the crossing point of the beams. The result is a very sharp spot of dose, or hot spot, which remains fixed at the isocentre of the piotron while the patient is translated within the bath. By moving the patient, it was possible to add 'dose spots' with individually optimised spot intensities to obtain a three-dimensionally shaped dose distribution."* [4]

There were numerous meetings debating the best technique to use: ring scanning or spot scanning? *"Strong emotions erupted at times,"* says Carl von Essen. *"I had my hands full as moderator."*

The research indicated advantages and disadvantages for both techniques. Eros Pedroni presented these in an internal report in November 1977. Spot scanning had been shown to provide an almost homogeneous biological effect (known as relative biological effectiveness, or RBE) across the target tissue; it required far less treatment time; and the dose delivery was significantly less sensitive to pion range

---

[4] The development of the water bolus was guided by Jaroslav Kohout, leader of the Mechanical Engineering Department of SIN.

errors. On the other hand, ring scanning was technically simpler, and applicable with lower pion ranges. It didn't involve water boluses and the complexity of moving a patient in three dimensions.

What was clear was that with both scanning techniques it was feasible to conform the dose to irregularly shaped targets. However, optimisation algorithms would be required in treatment planning. This was another field where the institute was later to take a globally leading role.

In early 1978, the project team moved into the new building and began to install the piotron facility. The spot scanning and ring scanning approaches were both pursued – for the time being.

Computers and treatment planning
The work on algorithms to plan patient treatment began. This was an ambitious and complex undertaking: the new treatment concept required thinking in three dimensions. It wasn't just a matter of calculating one beam's positioning and dose: the new dynamic concept involved coordinating 60 beams and the movement of the patient to target the dose accurately throughout the tumour. The task required a powerful computer, but this was a precious thing in 1978. It was a relief when the pion therapy project was assigned a PDP-11/45 computer, previously used as the central "batch" computer for the whole of SIN. But even with this, the programming process was painstaking.

"Just a few years before, scientific programming at SIN was performed with punched cards and by sending the code via radio waves to ETH Zürich," says Eros Pedroni. *"So at that time, the possibility of developing a complex medical application using a dedicated computer was for me very challenging but also very exciting. The complexity of programming for a dynamic therapy with identical concentric beams was at the very edge of feasibility with the computer power we had. The PDP-11/45 had only 32K of floating-point virtual memory, so treatment planning was broken down into a series of separate tasks and then you had to save each of those to magnetic tapes.*

"I was really astonished when we got permission from the director to buy a colour display for the computer. People would laugh at its quality today, but it was essential for us to do treatment planning at the time."

The ambitious goal was to deliver a homogeneous dose within the three-dimensional target, shaping the dose to fall off exactly in the tumour shape. But the ambition paid off. It was to result in probably the first treatment planning system in the world to make three-dimensional dose calculations from three-dimensional CT scan data.

Defending the project

As plans for pion therapy at SIN developed, so international interest in the project grew. This wasn't always positive. In February 1978, Carl von Essen and Jean-Pierre Blaser jointly wrote a letter to *Physics Bulletin*, defending the project against a claim from an English physicist that it was "socially irresponsible".[5] The physicist had expressed concern that the cost of such treatment would always be so great that it would never be widely available.

"The expectation is not that pion therapy may become a generalised cancer treatment," wrote von Essen and Blaser in response. "On the contrary, if it is successful at all, one may hope that for some types of tumours and anatomic distributions the probability of cure may be significantly improved, perhaps even in cases where it is now close to zero. The possibility of 'boosting', i.e. combination with classical treatment, is also considered. In such schemes, the cost of a 'small' pion machine would by no means be prohibitive.

"At present, the SIN project is a research programme only, aimed at establishing, maybe by 1985, what the merits of pion therapy really are. This can be done with an existing accelerator facility at costs comparable to those of one of the larger physics experiments in progress at such centres. Not to undertake it would be socially irresponsible indeed."

Activation

On 22 June 1980, at 4 a.m., the first beam was sent through the piotron. It was a momentous day and is recorded in an image of the first beam taken on radiographic film showing the dose ring for two different energies. The picture was signed by many of the people involved in the development of the machine.

Immediately after activation, issues arose that influenced the way forward. First, there was a problem with the coils, which were heating up and losing their super-

---

[5] J. P. Blaser and C. F. von Essen, 'Social irresponsibility', *Physics Bulletin*, 29 (1978), 55–56.

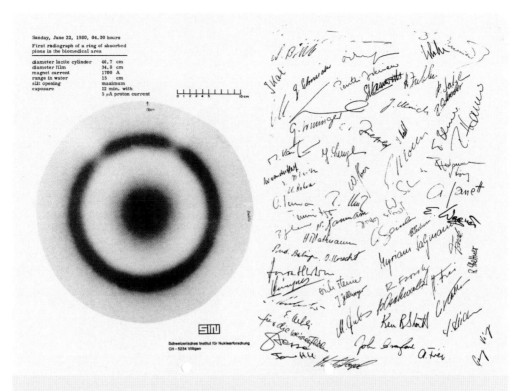

Figure 6. Many of the people involved in designing and constructing the piotron signed the radiograph of the first beam that passed through it.

conducting qualities. The source of the problem seemed to be neutrons (released, it was suspected later, from boron-containing glue within the coils). This needed to be addressed.

More significantly for the long term, the first tests on the piotron indicated that ring scanning would be impractical. The dose rate that the superconducting coils were capable of delivering was lower than expected – meaning that ring scanning would take far too long for most patients. So spot scanning became the favoured system.

"The dose rate limit was in the end the reason why the ring scan method was never used for treating patients, except maybe for the very first irradiations of skin metastasis," says Eros Pedroni. *"If we had decided in 1978 to start only with the ring scan technique and not develop the spot scanning method in parallel, the lifetime of the piotron would have been much shorter and the history of charged particle therapy in Switzerland could have been a very different one.*

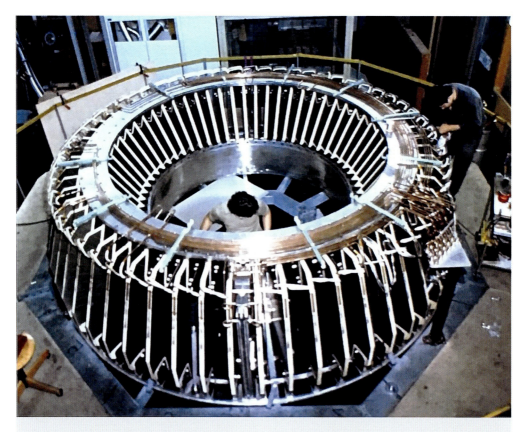

Figure 7. One of the piotron's huge torus superconducting coils is prepared for mounting onto the machine.

"The spot scanning idea soon became the main motivation for developing proton therapy with pencil beam scanning for deep-seated tumours, and scanning became the 'company brand' for successive medical developments at SIN and PSI."

A group led by engineer Ivo Jirousek perfected control and safety systems for the piotron, which were later further developed for proton therapy systems at PSI. Charles Perret continued to work on positioning the patient – a complex process involving casts, tissue-equivalent boluses, CT scans, X-rays and simulations. It was all necessary to ensure that the pion beams hit exactly the right part of the tumour, at exactly the right intensity, during each fraction (treatment session).

The group needed to test how their theoretical treatments might actually work on a human body composed of a variety of tissue types with different densities. So they constructed patient "phantoms" – dummies realistically simulating anatomy. In these, different mixtures of epoxy resins reproduced the varying densities and chemical compositions of muscle, fat and bones.

Eros Pedroni: *"We were very happy in the commissioning phase, after performing dosimetric experiments with the real beam and anthropomorphic phantoms and data from in-vivo dosimetry, to learn that the spot scanning method was sufficiently insensitive to body heterogeneities to avoid us having to compensate using a bolus."*

These were truly groundbreaking innovations, says Carl von Essen.

"Essentially, we were treating a cylinder composed of the patient, the material packed around them and the containing couch. The target volume within this cylinder, the tumour, was delineated after CT scanning at the nearby Kantonsspital Aarau. The CT data were reviewed by the radiotherapist, the target volume traced with interaction from the physicists, and the treatment planning started. This step already represented a major advancement in radiation therapy technology: the creation of complex three-dimensional target volumes from CT data and the programming of the planned volumes into the actual treatment."

It was an important time for the development of particle therapy generally. A year earlier, the first patients had been irradiated with pions with a single pion beam at Canada's particle accelerator, TRIUMF. It may not have been the space race, but each centre was highly conscious of progress being made elsewhere.

Hans Blattmann had visited the particle therapy facilities at Berkeley and Los Alamos. *"I saw a little bit of what others were working at, and from that took some ideas about how we could approach things at SIN. That was actually something that took some time before the first patients could be treated."*

A study of the radiobiology of pions in experimental animals began under the direction of Blattmann, in collaboration with the pathology department at the University of Bern.

"The first patients were naturally enough animals, and this also took some time. We started with small animals, but also pigs with areas that needed to be irradiated. There was always the question of whether it would be precise enough, whether the dose would extend outside the volume we wanted to irradiate, and if so whether it was something that could be tolerated.

"We discussed these issues very widely with many experts, and some of them provided us with valuable advice on what we could do to reduce the impact on the area surrounding the volume."

Hans Blattmann, Miriam Salzmann and Eros Pedroni began testing dosimetry and relative biological effectiveness (RBE) compared to conventional radiotherapy with photons.

Then, with testing complete, the first human treatments began. Carl von Essen remembers well the first people who volunteered for treatment. All were seriously ill, and pion therapy was a last resort.

"One was a Swiss lady from St Gallen suffering from metastatic melanoma with numerous cutaneous metastases. I treated these with a range of doses and measured both tumour effect and skin reactions. This enabled us to gauge the tissue tolerance, which was crucial to proceeding with dose planning. I eventually visited the patient, a lovely and patient woman, in her home to continue to measure and

Figure 8. The piotron. The patient cylinder on a movable transporter is on the left. It fitted within the piotron tube on the right. Around the piotron tube, the black asymmetrical water bolus can be seen, allowing movement of the cylinder in all directions within the piotron. Picture: Paul Scherrer Institute

photograph the late pion effects. She eventually died of melanoma and is remembered as a brave, selfless contributor to the development of pions for cancer therapy."

The early results were promising and attention turned to planning a full programme of patient research. The project's Medical Advisory Committee decided on three phases. Phase one would involve the treatment of patients with locally advanced and incurable cancer which had not yet spread, to find out more about treating various sites and gain knowledge on the dose effect on cancerous and normal tissue. Carl von Essen already had experience from Los Alamos of establishing effective dosages without severe side effects, at least in the short term.

Phase two was to focus on those cancers that were notoriously difficult to treat and where patient survival was currently very poor. Primarily, it would examine the effects of different pion doses on patients with bladder cancer. This followed a recommendation from a world authority on bladder cancer, given poor results to date treating locally advanced bladder cancer with conventional radiotherapy. Glioblastomas and pancreatic cancers – among the worst cancers in terms of curability and survival – were also to be examined in this phase.

Phase three would treat patients with earlier, more curable, cancers, comparing results with current conventional therapies.

In view of what was to come, Carl von Essen looks back at the decision to study bladder cancer with regret. But at the time, excitement was growing about the prospect of stopping difficult-to-treat cancers with such a novel technique – not just at SIN but internationally.

In a paper presented to the International Congress of Radiology in July 1981, von Essen, Blattmann, Pedroni and Perret wrote: "The initial responses and reactions are favorable and confirm the feasibility and accuracy of dynamic pion therapy." [6]

In the same month, Jean-Pierre Blaser and Carl von Essen were presented with the Swiss Cancer League's 1981 Award for Research in Cancer, as the promoters and

---

[6] C. F. Von Essen, H. Blattmann, J. F. Crawford, P. Fessenden, E. Pedroni, C. Perret, M. Salzmann, K. Shortt, E. Walder, 'The piotron: initial performance, preparation and experience with pion therapy', *International Journal of Radiation Oncology, Biology, Physics*, 8 (1982), 1499–1509, https://doi.org/10.1016/0360-3016(82)90609-5.

initiators of this new kind of radiation treatment. It was Blaser, said the commendation, who had had the idea of promoting the study of the biological reactions of pions and designed the technical facilities required for their medical use.

International interest ... and an explosion

Patient trials began in 1982. In the SIN annual report, Jean-Pierre Blaser reported that the process leading up to them had been "amazingly trouble-free".

SIN was now the only centre in the world providing high energy particle irradiation using a circular system of focused pion beams. Physicians were impressed with the unique three-dimensional spot-scan technique, planned using CT scans, and there was a growing queue of people wanting to see the piotron project at first hand. Many medical doctors from Switzerland and abroad wanted to take their sabbaticals at SIN.

There was particular interest from Japan, where trials were continuing apace in particle therapy at the National Institute of Radiological Sciences, Chiba, and the Particle Radiation Medical Science Center at the University of Tsukuba. Japanese scientists had a special interest in pions, because their existence had been predicted by Japanese theoretical physicist Hideki Yukawa in 1935, 12 years before they were actually discovered.

Carl von Essen remembers how a series of Japanese physicists and radiotherapists came to work with the piotron team. The visits continued over many years, supported by funding from a Japanese newspaper.

"We forged long-lasting bonds of friendship," he says. *"Many other physicists also visited, including Peter Fessenden from Stanford University and Ken Shortt from the University of British Columbia, Vancouver. It was a truly cosmopolitan group. Naturally, English was the main language in many of the meetings, while I struggled with my German in others."*

In turn, both Carl von Essen and Eros Pedroni travelled to make presentations abroad in response to growing interest. During the years of the piotron project, Pedroni visited particle therapy projects in the United States, Canada and Japan and met fellow physicists doing similar work at international meetings.

Figure 9. Carl von Essen with visiting physician Jun-Etzu Mizoe from Japan (centre) and radiation oncologist Gerd Bodendoerfer.

Such international contact-making was sometimes made easier by the technical vagaries of the complex piotron. On one occasion, the turbine of a vacuum pump exploded – sending metallic fragments through the piotron tank and damaging the superinsulation foils which maintain ultra-low temperatures around the coils. The piotron was shut down and it took two months to repair the damage, but eventually the piotron programme resumed.

Eros Pedroni remembers it as the only accident involving the piotron – there were certainly none involving patients. But such technical down time and the long winter shutdowns of the accelerator gave von Essen a chance to network. Based, as he was, at an isolated science facility in the middle of the Swiss countryside, it was important to develop contacts through personal meetings with Swiss medical societies, European cancer organisations and individual hospitals.

"I took up that opportunity to travel around the country and around Europe to help recruit doctors, and ultimately get patients into our programme," says von Essen.

"I travelled extensively with the Swiss Railways and really saw and enjoyed Switzerland. I also travelled to Italy, Austria, Germany, France and the Netherlands for medical networking. I eventually got a teaching appointment at Basel University, as Ausserordentlicher Professor, and taught a course – in German – in radiobiology. I also attended clinics, examined patients and observed operations and cystoscopies."

The patient trials

Phase one of the clinical programme began simply enough, with static treatments of multiple skin nodules. The main purpose was to allow clinicians and technicians to better understand the biological effectiveness of this new type of irradiation compared to standard photon (X-ray) radiotherapy.

The next step was to start "dynamic treatment", moving the patient around in three dimensions to treat larger, metastatic tumours close to the surface of the body.

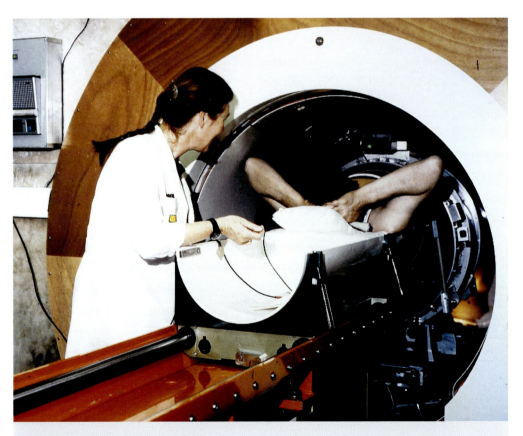

Figure 10. A patient is eased into the piotron, having been carefully positioned within a treatment "cylinder". Picture: Paul Scherrer Institute

These surface tumours were less complicated to treat than tumours deep inside the body because different tissue densities inside the patient's body did not have to be taken into account.

Patients were given treatments of 15 to 40 minutes in 20 fractions. Conducting the treatments were the medical staff of Carl von Essen, two nurses and radiation oncologist Gerd Bodendoerfer. The piotron facility was now a fully operational clinical unit, and for the first time the SIN organisational chart included a department marked "Medicine".

But it was clear that the shift towards clinical work at a scientific institution required more than new staff and equipment. Now that people who were ill and vulnerable were regular visitors, there was a need to address the psychological side of health care.

"The staff at SIN, the epitome of a purely scientific and intellectual facility, was now seeing for the first time human suffering and despair among patients undergoing rigorous experimental treatment protocols," says Carl von Essen. *"Both patients and staff were facing unaccustomed stressful experiences.*

"I instituted a weekly therapy session with a Zürich psychologist, who was experienced with cancer patients. We discussed not only the patients' problems but our own in addressing interactions with patients. The sessions were very beneficial and well attended. Another innovation was a retreat for the entire medical section to a ski resort, where morning sessions of brainstorming organisational and technical problems were followed by cross-country skiing and finally relaxing banquets."

With phase one complete, phase two began, starting with dynamic treatments of deep-seated tumours in the pelvic region. Even early on, there was some favourable evidence of tumour control. But that was not the whole story. It soon became clear that there were also unpleasant side effects associated with the treatment.

"This was particularly seen in patients with cancer of the bladder," says von Essen. *"I attended the university hospital in Bern, along with my collaborating urologist, Dr Urs Studer, to examine and cystoscope the patients, and saw the disturbing results."*

Their tumours had gone, but in some bladder cancer patients the radiation had also caused fibrosis, which in turn caused painful contractures. The patients had to be given urgent surgery to ease their symptoms. Carl von Essen and Urs Studer reported on the experience in a paper published in the journal Cancer in 1985.[7] Of 36 patients, 21 were cleared of their bladder cancer after treatment. But of the 11 patients who received higher doses of pion radiation, seven had late and severe side effects.

"We were all depressed by this situation and it became clear that pions did not have a role in the treatment of bladder cancer, at least not with the existing conditions of dose, fractionation and volume," says von Essen.

There were now fears that pion therapy was not the "magic bullet" that physicists, radiobiologists and clinicians had hoped for. But these were early results. The work on bladder cancer ended in 1984 but continued on other difficult-to-treat tumours: reviews took place, doses were recalculated and research progressed.

Departures
Carl von Essen and his team tried to work out what was happening to bladder cancer patients, comparing their data with information coming out of the pion projects in Los Alamos and Vancouver. There was an issue with controlling the dose distribution of pions and concentrating the effects of their high LET energy at the specific points being targeted. The bladder was especially vulnerable to the effect of this extended dose. The problem seemed to be due to the inherent physical characteristics of the pion.

Carl von Essen acknowledges that he began to feel increasing disenchantment with the pion project – not just because of the disappointing early clinical results, but because he felt out of place at SIN. In his autobiography, *Memory is a Dwelling Place*, he says he was always known at SIN as the "hippy American doctor".[8] He certainly felt at odds with the climate of physics and engineering at SIN.

---

[7] U. E. Studer, C. F. Von Essen, J.-B. Enderli, G. Bodendörfer, E. J. Zingg, 'Preliminary results of a Phase I/II study with pi-meson (pion) treatment for bladder cancer', *Cancer*, 56 (1985), 1943–52, https://doi.org/10.1002/1097-0142(19851015)56:8<1943::AID-CNCR2820560809>3.0.CO;2-B.

[8] Carl von Essen, *Memory is a Dwelling Place* (Shyamal Books, 2008).

In 1983, seven years after his arrival, he was offered a job as a World Health Organization Consultant in Clinical Oncology to Sri Lanka, and decided to accept. Aware that he was leaving the pion programme at an important point – a new research programme on the long-term effect of pions on animal tissues was just beginning – he helped find a successor, Richard Greiner, in 1984. Greiner took over from von Essen in early 1985.

Carl von Essen was not the only one to leave the pion project. His forthcoming departure seemed a major juncture for Eros Pedroni too.

"The technology of the piotron was settled, Carl von Essen was leaving and the results of the bladder treatments were disappointing," says Eros Pedroni. *"In addition, biologists were anticipating the end of radiotherapy – they were talking about new interferon drugs solving the cancer problem instead. So I thought I would like to do physics research again – the institute at the time was still mainly focused on basic research. For me, the pion therapy project was a nice experience, but I decided to leave the project in 1983 and returned to SIN's physics department."*

But it was by no means an end. In fact, it was just a beginning. The pion project may have been experiencing setbacks, but it had already laid the groundwork for all the important developments in particle therapy that followed at PSI. The piotron was probably the first radiation therapy treatment system in the world that scanned in three dimensions, making three-dimensional dose calculations from CT data.

Eros Pedroni would be back to perfect these systems of treatment delivery which were to spread around the world. And, in another part of SIN, a separate programme was demonstrating exactly what could be achieved by using a particle heavier than the pion, one that would be at the centre of future developments: the proton.

Chapter 3

1982–1992
Protons for eye tumours

In June 1982, Professor Leonidas Zografos, consultant ophthalmologist at the Jules Gonin Eye Hospital, Lausanne – received a phone call out of the blue from Charles Perret.

"He asked me if I knew anything about the proton beam irradiation of uveal melanomas," says Zografos. *"I told him that of course I knew something about it from the literature, and he asked me if I was interested in being involved in a treatment project with proton beam irradiation."*

Physicist Charles Perret had first become fascinated by the potential of proton beam irradiation during a visit to the treatment facility of the Harvard Cyclotron Laboratory, Boston, in the 1970s. The Boston facility had been the first in the world to treat melanomas of the eye with protons in 1975, and its early trials, run in collaboration with the Massachusetts Eye and Ear Infirmary, proved successful.

Perret was keen to pursue a similar project at SIN but, like Harvard, he needed a collaborating hospital to provide clinical expertise – and patients. There was an obvious candidate in Switzerland. The Lausanne University Eye Clinic at the Jules Gonin Eye Hospital had pioneered the use of radioactive "applicators" (applying a radioactive source close to the cancer, also known as "brachytherapy") for eye tumours in Europe. Led by Zografos and Professor Claude Gailloud, the clinic was the only ocular oncology centre in Switzerland.

"I remember the phone call from M. Perret well," says Zografos. *"He asked me some details about my experience. It was a Monday, and I told him that on Thursday we were holding a conference in our auditorium here in Lausanne for residents and local ophthalmologists about the use of cobalt-60 applicators for uveal melanomas. I said come along and then we can discuss it.*

"He came to Lausanne, attended the conference and told me that it was the first time that he had heard a scientific report from an ophthalmologist: he was a phys-

icist, of course, not a medic. He told me all about the intention to create a proton beam irradiation project in Villigen, and I told him I was very interested."

It was an exciting prospect. Tumours of the eye are rare, but they cause loss of vision and can fatally spread (metastasise) to other parts of the body. Ocular melanomas are the most common eye cancers – affecting around six in every million people. Most cases start in the area of the eye called the uvea (the pigmented middle layer of tissues forming the eyeball). Because of the risk of metastasis, early and effective treatment is critical.

"Until the 1950s and 1960s, enucleation [eyeball removal] was considered the standard therapeutic approach for uveal melanomas," says Zografos. *"However, the limited efficacy of enucleation in preventing metastatic spread of intraocular melanomas, observed by various investigators, as well as the first positive therapeutic results using conservative management techniques [preserving the eye], created an increasing interest in new techniques.*

"In Switzerland, the first radioactive applicators, Cobalt-60 applicators of Stallard, were introduced in 1969 in Lausanne. The use of these radioactive applicators allowed the accumulation of expertise in ocular oncology at the University Eye Clinic of Lausanne. Recruitment of patients increased year after year. In the late 1970s and early 1980s, we had added additional therapeutic modalities, including Ruthenium applicators and surgical removal of intraocular tumours, to our therapeutic armament."

A month after his phone call, Perret wrote to Zografos confirming his intention to start a project for the proton beam irradiation of uveal melanomas. On 3 September, Perret elaborated on his plans in an official memo to key SIN staff, including Jean-Pierre Blaser, Wilfred Hirt (the deputy director), Carl von Essen, Hans Blattmann and Eros Pedroni. By now, the plans had a name: Ophthalmological Proton Therapy Installation SIN, or OPTIS.

Perret pointed out that survival after radiotherapy seemed to be just as good as after enucleation, and that irradiation with protons – as developed in Boston and currently also being trialled in Moscow – was "the ideal treatment method".

SIN	KATEGORIE: PROTONEN-PROJEKT	ORDNUNGS-NR. AN-70-01	SEITE NR. 1	von TOTAL 20±Anlage
TITEL: STRAHLENTHERAPIE AN MELANOMEN DER ADERHAUT MIT PROTONEN VON 72 MeV		NAME: Ch. Perret/HCM		
		DATUM: 3. September 1982		

Verteiler:

Prof. J.P. Blaser
Dr.   W. Hirt
Prof. C. von Essen
Dr.   P. Schwaller
Dr.   U. Schryber
Dr.   C. Tschalär
Dr.   J. Domingo
Dr.   R. Hoffmann
Prof. H.J. Gerber

Hr. R. Balsiger
Dr. T. Stammbach
Dr. S. Jaccard
Dr. H. Blattmann
Dr. E. Pedroni
Hr. F. Adamec
Hr. J. Ulrich
Hr. C. Perret   (3)
Bibliothek   (3)

Referenzfile Dok./Inf.
Originale: M. Seiler

STRAHLENTHERAPIE VON MELANOMEN DER ADERHAUT MIT PROTONEN VON 72 MeV

Projekt OPTIS

1. Begründung
2. Protonentherapie der Augenmelanome in Boston
3. Therapie mit Helium-Ionen in Berkeley
4. SIN-Anlage OPTIS
5. Betriebsbedingungen bis Ende 1984
6. Betriebsbedingungen ab Ende 1984
7. Zeitplan und Organisation

Anlage:
Auszug aus Arch Ophthalmol - Vol 100, June 1982
"Proton Beam Irradiation of Uveal Melanomas"

Anhang:
Einige wichtige Daten.

Figure 11. The memorandum circulated by Charles Perret in 1982 setting out plans for Project OPTIS.

"Radiation therapy has the advantage that the affected eye can normally be saved, usually without increasing visual loss," he wrote. *"If proton radiation therapy were convenient to use, almost all melanoma patients could be treated with it . . . The cost of the project is 350,000 Swiss francs. It is possible to treat the first patient by the end of 1983.*

"The group in Boston supports the introduction of their method in Europe because the capacity of their facility (60 patients per year) is almost full. The Steering Committee for Pion Therapy at SIN has recommended the introduction of proton therapy of eye melanoma at SIN. Prof Gailloud, whose ophthalmic clinic in Lausanne has a central position in this field in Switzerland and 40–50 cases of ophthalmic melanoma from Switzerland and abroad each year, has declared his clinic's willingness to participate significantly in the project."

SIN's 72 MeV proton beam from the Injector 1 cyclotron was well suited for the purpose, said Perret: it was the precise energy level needed to penetrate an eyeball and stop at a tumour on the retina. Essentially, SIN was to build an optimised copy of the Boston facility, and provide proton therapy for eye tumours for the first time in Europe.

Projects in Louvain-la-Neuve (Belgium) and Uppsala (Sweden) were also in preparation at the time, so this was an important and potentially prestigious project for SIN. Things moved fast. On 1 October 1982, Jean-Pierre Blaser wrote to Professor Gailloud in Lausanne, confirming that the directors of SIN had formally approved the OPTIS project.

Building work, said Blaser, would be completed in 1983, with the aim of treating the first patient by the end of that year. During 1984, the number of patients would be restricted because of the need to complete and perfect the installation, as well as the limited time the proton beam would be available for therapy purposes. But the capacity of the facility would increase by 1985.

"During the commissioning phases, close collaboration between the SIN group and your institute seems to me to be one of the conditions for the rapid success of the project," wrote Blaser.

That close collaboration continues to this day.

Working across borders

Physicians from both the Lausanne University Eye Clinic and SIN immediately recognised how important it was to establish awareness of the project across Europe if they were to receive patient referrals for such rare cancers. So while Blaser was exchanging letters with Lausanne confirming the collaboration, Gailloud, Zografos, Perret and von Essen were in Geneva meeting representatives of the Union for International Cancer Control (UICC), an organisation founded in 1933 to encourage the sharing of cancer knowledge, skills and technologies globally.

On 5 October 1982, they presented the OPTIS project to the UICC ophthalmological section, including evidence gathered from the Boston project.

"We received unanimous support from the organisation for the application of this therapeutic modality in European patients," says Zografos. *"It was a very important turning point, to gain the support of the European oncological community so early."*

UICC agreed to financially support SIN staff internships in Boston, as well as visitors from the rest of Europe to the SIN facility.

Charles Perret, the driving force behind the project, was put in charge of creating the new facility, with Samuel Jaccard his co-manager. But making the project a reality was not simply a matter of extending the beam line and constructing a new facility. During 1983, skills were developed, networks created and technologies perfected – all of which would be instrumental to the success of the OPTIS project.

One key initiative built on the success of the UICC meeting and placed the SIN/Lausanne University Eye Clinic collaboration at the centre of European ocular oncology. Zografos and Gailloud organised the first European course on the diagnosis and treatment of intraocular tumours in Nyon in Switzerland in April 1983. The main speakers were Carl von Essen, Charles Perret, Leonidas Zografos and Evangelos Gragoudas – the man who had pioneered the use of proton therapy for the treatment of eye tumours in Boston. He was, like Zografos, Greek by extraction, and the two formed a bond of friendship that served cross-Atlantic collaboration well.

"This was an extremely important meeting, because after this point we started to receive patient referrals, and so we had to put in place the systems and facilities to

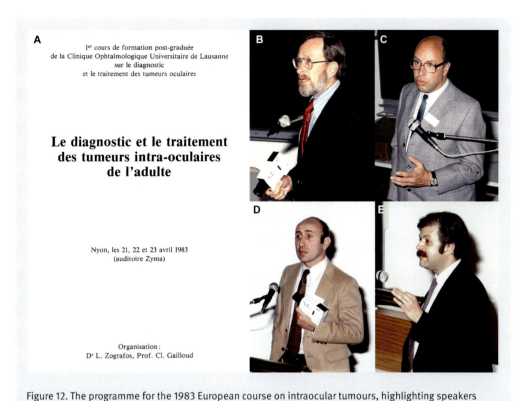

Figure 12. The programme for the 1983 European course on intraocular tumours, highlighting speakers (clockwise) Carl von Essen, Charles Perret, Leonidas Zografos and Evangelos Gragoudas.

be ready for those patients," says Zografos. *"It also allowed the creation of a collaborative group in Europe."*

The willingness of Harvard University and the Massachusetts General Hospital (MGH) to share their knowhow with the European initiative was a key factor in helping the OPTIS project to advance so quickly.

Charles Perret, during his trips to Boston, learned much from Professor Michael Goitein – the physicist who had developed three-dimensional treatment planning systems and image-guided beam delivery for radiation therapy at MGH. These new technologies had made precise proton therapy possible. Alongside this, Goitein had developed treatment planning software specifically for the proton beam irradiation of ocular tumours. It was called EYEPLAN and he gave Perret a copy.

Perret took EYEPLAN and adapted it for use at PSI, formulating a dose and fractionation regime using radiation measurement equipment donated by MGH.

About EYEPLAN

The EYEPLAN treatment planning software for ocular melanoma patients was developed at Massachusetts General Hospital (MGH) by Michael Goitein and T. Miller in the late 1970s and early 1980s, and it is still used today. It arrived at SIN through Charles Perret's collaboration with MGH, and PSI's version was in turn shared with the Clatterbridge Cancer Centre in the UK when it began proton treatment for uveal melanomas in 1989. Clatterbridge developed the software further, rewriting the code so that it could be used on PCs, and adding eyelid modelling and iris tumour and clipless planning options. Proton therapy centres around the world have obtained copies of EYEPLAN and updates from Clatterbridge.

In the early 2000s, PSI created a hemispherical eye volume option for EYEPLAN which became available to all users. Today, EYEPLAN ownership is retained by MGH, PSI and Clatterbridge.

Zografos also did his own research on preparing patients for proton irradiation. Before sending patients to SIN for proton therapy, he would have to conduct an intricate procedure to accurately mark the tumour-affected areas in need of irradiation. The technique, developed by Professor Gragoudas and used to this day, involved attaching tiny clips made of the rare metal tantalum to the outside of the rear portion of the eyeball to delineate the tumour outline. The clips appear on X-rays as black dots and are used by medical physicists to define the exact area of the tumour – which cannot otherwise be seen on X-rays.

Zografos worked on these techniques alongside Gragoudas and his colleagues in Boston throughout September.

"I learned about tantalum clip fixation, patient follow-up and the whole procedure. I got to meet Professor Goitein, who devised the treatment planning programme, and also my good friend Gragoudas. Then I came back and started visiting Villigen two or three days every month, to discuss all the points with Perret."

Perret and the chair

One of Perret's main concerns – both during the project set-up and throughout his time on OPTIS – was to perfect the chair in which the patient was to receive proton treatment. This could be no ordinary chair. The patient had to be seated directly in front of the proton beam outlet and highly accurately and stably positioned, or else treatment might miss its target – causing treatment failure or damage. Ocular melanomas can be just a few millimetres in size, so even the smallest twitch from the patient might cause the protons to hit the wrong area.

The procedure developed was that the patient's head would be completely immobilised using a custom-made mask and a bite block cast to fit the patient's teeth. These two components were fastened to a mask frame, which was in turn fixed to the chair – the result being that, when the patient was ready for treatment, only their eyes could move. The chair had to be positioned at extremely precise heights and distances from the emission point of the proton beam.

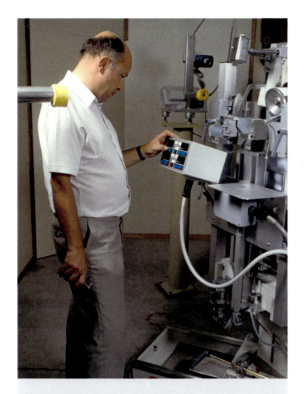

Figure 13. Charles Perret adjusting the treatment parameters of the stereotactic chair.
Picture: Jules Gonin Eye Hospital

All these issues were of concern to Perret, who in 1983 turned his skills to improving the prototype chair design provided by Massachusetts General Hospital.

"Charles Perret built the facility with very few resources," says Emmanuel Egger, who worked closely with Perret from 1987 onwards and was to succeed him in leading OPTIS in 1989. *"The chair was his biggest project, along with the steering electronics for the proton beam. At the beginning it was not a very comfortable experience for patients."*

The chair that Perret developed, known as a "stereotactic chair", enabled treatment through highly accurate positioning. It was movable in all three dimensions and could be positioned with a precision within 0.1 mm. In all, it took Perret three years to perfect. With time, it became famous. Perret would give a copy of his drawings to the group building an ocular tumour proton treatment facility in Clatterbridge in the UK.

"The chair was very influential, and several other institutions wanted to buy a copy," says Emmanuel Egger. *"But it was decided that SIN was not a chair factory, and that we should sell just one copy, to Loma Linda University Medical Center in the United States, which had already put in an order."*

Later, in the early 1990s, PSI was to license the OPTIS chair technology to a Swiss engineering company which manufactured similar chairs for proton therapy centres in Germany, France, Italy, the USA and Asia.

Throughout 1983, Perret and his team worked on patient immobilisation and chair construction. But they also had vital work to complete on dosimetry – measuring how much radiation is absorbed by tissue and calculating the optimal dose delivery. Zografos remembers that, amid the excitement of being involved in such an innovative project, there were still many daunting unknowns.

"I have to say that in the 1970s to early 1980s, the first question was: is it safe to conduct conservative treatment on eyes with melanoma, or should we always

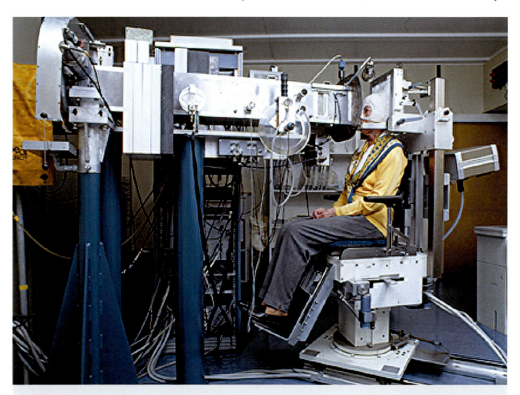

Figure 14. The mechanical stereotactic chair, as completed by Perret and used until the OPTIS2 redevelopment. Picture: Jules Gonin Eye Hospital

remove? The second question was: if we treat with irradiation treatment, what is the correct dose to cure this type of melanoma? We didn't have a clear idea, and radiotherapists were unable to answer the question. For cobalt treatment I fixed the dose at 90 Gy and later other researchers arrived at the same figure. We were in agreement. But for proton beam irradiation there was no consensus on the best dose."

Treatment begins

The first patients were treated in March 1984, just three months later than Perret had forecast, and 100,000 Swiss francs below the predicted budget. One of the first patients was a photographer from a popular French-language newspaper in Switzerland, and with a colleague he put together an article published just days

Figure 15. The article in the Swiss newspaper *La Tribune le Matin* (or Tlm) that first drew national attention to the groundbreaking OPTIS project at PSI.

after his treatment began. The headline was "A European first: accelerated protons for eye disease". The article, by Véronique Tissières, began:

"This left eye, how many times has Jean closed it during his long career as a photographer? 'It's a bit like a child who is neglected and who rebels,' he explains symbolically. The first signs of this rebellion: a grey veil blurs his vision. Last autumn, he underwent an examination which revealed the presence of an intraocular tumour located near the optic nerve. Jean is the first patient to follow a treatment unique in Europe, set up by a medical team from the Lausanne ophthalmological clinic, in close collaboration with physicists . . . During four sessions of a few seconds each, the intraocular tumour was treated by irradiation of accelerated protons."

Only around 20 patients could be treated in the first year. SIN's other research work also required the beam from the injector cyclotron, and the beam line could only be diverted to the OPTIS facility for a few hours a day, one week per month. Radiotherapy had to be performed at the end of the day or during the night.

"The conditions were quite difficult," says Zografos. *"In the week that we had the beam, it was allocated to us after 6 p.m. To split the beam and take it to the OPTIS facility usually took three to four hours, so during the first year we started treatment of the patients around ten at night, and usually we were finished by midnight."*

Nevertheless, patients came – and kept on coming. By the second year, numbers had increased to 100. Though the treatment was experimental, there was the prospect of saving their sight – and possibly their lives.

Zografos: *"The patients were really happy to have this treatment. They came from France, Belgium, Holland, England, Spain, Italy, Greece – and the ophthalmologists who sent the patients said such good things about our work and the whole procedure, so the patients arrived enthusiastic to be treated. I think that, being treated in an experimental physics institute, they had this feeling of being like astronauts! The whole ambience at the time was not medical: it was really like an engineering atmosphere."*

Having completed all the preparatory consultations and procedures at the Jules Gonin Hospital in Lausanne, Zografos attended all the early proton treatment sessions at SIN as well: the treatments were new and involved, so close consultation

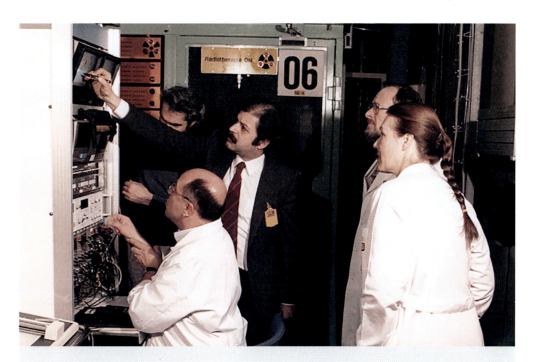

Figure 16. The OPTIS team from SIN and Lausanne gather around the monitoring station during the treatment of the first patient in 1984. From left to right, Samuel Jaccard, Charles Perret (seated), Leonidas Zografos, Carl von Essen and therapeutic nurse Teresa Hirt. Picture: Jules Gonin Eye Hospital

between team members was essential. They also needed to devote time to perfecting dosimetry and technology.

"It was a very friendly atmosphere, and I have excellent memories," says Zografos. *"The years between 1984 and 1986 were the years of technical development. I used to go to Villigen two days every month to discuss medical problems and technical developments. The main developments were made by Charles Perret. He was a genius – a very inventive physicist. Usually the procedure was that I had a discussion with Perret and made some proposals. He said okay, and one or two months later he would come up with the solution.*

"It was really a great pleasure for me to work with him because he was able to understand all my medical problems. Proton beam facilities are not the same around the whole world, and each one is special. We could not use the parameters of another facility: we had to adapt ours for the treatment of our patients. So we redeveloped everything: technical developments for the beam, the coils, the contra-indications, the fixations for the eyes, the multi-modular collimators [devices which narrow a beam of particles] *– everything."*

As their experience of proton therapy grew, Zografos, Perret and von Essen fine-tuned the optimal doses.

Zografos: *"We started with the theoretical dose, and quickly learned that it was excessive. During the first year we reduced the dose from 70 to 60 Gy, and then we made further adaptations for our own proton beam facility. We had long, long discussions, always between von Essen, Perret and myself, and decided to progressively reduce the dose. And then we conducted a trial to fix the best dose, and that's what we continued with."*

All the while, the pion project ran alongside the nascent proton programme. They were separate ventures, but both Charles Perret on the OPTIS side and Hans Blattmann on the piotron side made a point of comparing notes.

"For a long time we shared an office," says Blattmann. *"Each of us did his own job, but we got interested in the progress of the other, and we had many discussions – not so much about scientific practicalities, but broader issues such as how we can find more people, and how we could spread interest in what we were doing."*

The future of the project reviewed

The OPTIS project was formally reviewed two years after its instigation. On 24 June 1986, Charles Perret presented a report on the past, present and future of OPTIS to Jean-Pierre Blaser, Richard Greiner (von Essen's successor as director of the medical programme at SIN) and other SIN representatives at a meeting in Bern. The question was: should the project stop, now that the development stage was completed?

Zografos: *"Here, after all, was an institute whose purpose was research and development. And now the development was finished, we either needed to end OPTIS or put the whole development within a medical structure at SIN. After the discussion we agreed that, although we had finished the main development, there remained additional developments to be made and we should continue at SIN. We took the decision to continue and provide additional financial support, and we never discussed this issue again. We have stayed the same way since 1986."*

Part of the reason that continuation of the project was possible was that, by 1986, more patients could be treated. This brought more income for SIN and a greater ability to carry out research and improve the service. A new small injector cyclotron, commissioned in 1984, came online at SIN in 1986, freeing up Injector 1 capacity for OPTIS. It could now be used for a full week in every month. In the other weeks, patients were prepared for irradiation in Lausanne and then, at SIN, had masks made and went through a simulation. This meant that the number of patients treated annually could increase to around 200.

The important role of the OPTIS project within the institute was now established. Its international reach extended, aided by the continuing annual courses on intraocular tumours organised by the Lausanne clinic. The fourth course, in April 1986, took place at SIN. In 1987 Zografos and his team also organised the Second International Congress of Intraocular Tumours, which further consolidated the position of the Swiss group in the international ocular oncology community.

As Andreas Pritzker, a manager at SIN and PSI in the 1980s, says of OPTIS in his history of the institute: "This totally unique project in Europe garnered strong international recognition and led to a great number of foreign patients, especially from Italy. The Swiss Cancer League, the Federal Office of Public Health and the health insurance companies assisted constructively and without creating complications in solving the unusual administrative questions regarding coverage for the therapy. Revenues increased accordingly and were a welcome increase in the research budget of SIN's medical program." [9]

A change of leadership
Part of the expansion plan was the addition of a new physicist to the OPTIS group. Emmanuel Egger joined in 1987, having previously worked as a software developer. He qualified as a medical physicist while at the institute.

"I had never heard of this treatment before applying for the job, but it sounded very interesting to me. I got the job and I liked it. That's why I worked there for 15 years."

Although OPTIS had moved on considerably in terms of optimising treatment since 1984, the experience for staff was not always easy. Egger's first year at the facility involved intensive learning, long hours and constant improvisation.

"When I joined SIN, Mr Perret and myself travelled to Lausanne regularly to assist at the surgical interventions for the positioning of the tantalum clips. This way, we

---

[9] Andreas Pritzker, *The Swiss Institute for Nuclear Research SIN* (Norderstedt: Books on Demand, 2014).

learned a common language between physicist and ophthalmologist. Every physicist or MTRA [medical-technical radiology assistant] working with the OPTIS group had to assist in several surgical interventions to learn the whole procedure – from surgical intervention, to moulding, treatment planning, simulation in the treatment position and treatment itself.

"When I began, the treatment planning software was running on the main computer of the institute. Every first Monday of the month was reserved for maintenance and the computer was not available, but unfortunately, this was often also the first day of our treatment week. As a consequence, we were not able to correct the treatment plans until the afternoon if the patient was not able to take the position we had calculated. So we regularly had to work late hours at these times. Another issue was that after a system update on the main computer our treatment planning software didn't work any more. So we had to spend time finding the cause of the problem and fixing it instead of working on the treatment plans."

And the working environment was still that of an austere experimental centre, not the comfortable clinical facility that exists today. Swiss winters could be harsh.

"At the beginning of my time, the steering electronics of the proton beam were in the main SIN hall. It was cold there in winter – and once there was a storm and I had the whole electronics underwater. We decided to make it a little more comfortable for the people who worked there and move a little bit inside the treatment room, so that we could heat it. I remember that Charles Perret never felt the cold. Even in the coldest of winters he was just wearing a shirt with short sleeves."

But Perret's energetic presence was not to be around for much longer. In 1988, he travelled to Boston for a six-month sabbatical, handing over the technical leadership of OPTIS to Egger. On his return to Switzerland, Perret moved to another PSI role outside the Radiation Medicine Division. He continued to work at PSI until his retirement. *"He was a man always interested in building up new projects, so the routine treatment of patients was not for him,"* says Emmanuel Egger.

There were other major changes. SIN merged with EIR (the Swiss Federal Institute for Reactor Research) to form the Paul Scherrer Institute (see Chapter 4). And in 1989, Richard Greiner departed, to be succeeded by radiation oncologist Gudrun Munkel (later Goitein), who took on overall responsibility for radiation treatments at PSI.

Emmanuel Egger succeeded Perret as OPTIS leader. The project was by this point treating around 200 patients annually as planned, and "running very well", according to Egger, who was in charge of treatment planning, dosimetry and patient treatment. A description of the patient assessment, preparation, treatment and follow-up procedures is given in the section below.

The OPTIS procedures

Patients referred to the University Eye Clinic in Lausanne were carefully evaluated there. This involved detailed drawing and photography of the fundus (back of the eye) and measurement of eye length and tumour thickness, shape and location using ultrasound and imaging. During the early years, the main indication for proton beam radiotherapy was tumours that could not be treated with brachytherapy, though this was refined and enlarged as time went by.

Patients then underwent a surgical procedure under anaesthesia to precisely mark the tumour. A shadow of the tumour base became visible on the back of the eyeball when light was shone through the eye, and Professor Zografos sutured three to seven tantalum clips onto the sclera (the outer surface of the eyeball) to mark the border.

Patients left hospital two days later and went to SIN/PSI, where a custom-moulded head-holder containing a bite block was built for immobilisation. Then the patient was installed in the positioning chair developed by PSI and had a simulation session to obtain data for treatment planning. X-ray pictures showing the tantalum clips in a beam coordinate frame were obtained while the patient's gaze was fixed on a small target light set at a known angle. The relative positions of the clips could be accurately measured from these X-rays. Treatment was planned using the EYEPLAN programme developed by Goitein and Miller at Massachusetts General Hospital and later modified by Perret for use at PSI.

Using the clip coordinates as measured on the simulation pictures and ultrasound measurements of eye length, EYEPLAN built a model of the eye, the tumour base and the tumour profile. With this model, an optimal treatment position could be calculated – one allowing complete irradiation of the tumour but with a safety margin (usually 2 mm), minimising irradiation of the optic disc and nerve, the macula, the ciliary body and lens. EYEPLAN also determined the penetration depth of the proton beam, the width of the Bragg peak and the shape of the aperture required to correspond to the tumour shape.

About collimators

Figure 17. A selection of collimators. Picture: Paul Scherrer Institute

Conventionally, radiotherapy is delivered through a single beam, shaped to the target tumour using a collimator – a metal device, often individualised for each patient, which aligns the radiation into a narrow beam in the shape of the tumour being treated. Collimators are used in X-ray (photon) radiation therapy using LIN-ACs, and are also used for proton therapy of the eye at PSI. The spot (pencil beam) scanning systems for deep-seated tumours developed at PSI (see Chapter 4) eliminated the need to manufacture and use individualised collimators.

Proton beam range was selected to place the optimal radiation dose at the tumour, with the collimator border defining the lateral fall-off of the radiation.

When the PSI staff had worked out the treatment plan, it was sent by fax to Professor Zografos and discussed during a telephone meeting and amended if necessary. Sometimes Zografos corrected the target volume, sending back an amended drawing via fax.

Simulations were carried out on Thursdays in the week before treatment. Then, on the Monday of the treatment week, the patient positioning was verified, with the chair and the light fixing the gaze adjusted until the eye was placed exactly as in the treatment plan. If this position could not be achieved, the treatment plan was modified according to what was possible for the patient. Then a file providing the data for the required copper collimator, shaped to the contours of the tumour in the treatment position, was sent to the PSI workshop and fed into a computer-driven milling machine.

Also on Mondays, before the treatment of every patient, a treatment simulation was completed on a plexiglass "phantom" in order to verify that the dose, the depth of the proton beam and its modulation had been correctly adjusted.

From Tuesday to Friday patients received their therapy. During this time, they lived at a hotel near the institute. Treatment was delivered in four fractions, usually on four consecutive days. Most of the patients were treated with a proton dose of 54.5 Gy. There was always an ophthalmologist from Lausanne at PSI during the treatment.

After treatment, patients were asked to return to the University Eye Clinic of Lausanne for follow-up examinations at six months, one and a half years, three years, five years, and then every two and a half years. If they were unable to return to Lausanne easily, follow-up data were collected by contacting the referring ophthalmologist. Follow-up examinations included ultrasound measurements of the tumour, fundus photography, measurement of visual acuity and ocular pressure, and diagnosis of possible complications.

Analysing outcomes
Egger had two broad priorities for OPTIS alongside the day-to-day responsibility of providing the best possible treatment to patients. One was to optimise the treatment process further so that more patients could be treated each year. The second was to work on the database of information about patient treatments and outcomes and provide an authoritative analysis of treatment success.

"We optimised treatment so that we could handle 240 patients per year with nearly half the personnel we used at the beginning," says Egger. *"I remember that at a conference I told a physicist from the Institut Curie in Orsay, France, that it took 15 minutes to treat each patient. She said she didn't believe me, so I invited her to come and see how we worked. She did, and she was very surprised how we could treat so many patients. She saw we were a very good team: there were generally only two people treating each patient. Each had their task to perform and knew that*

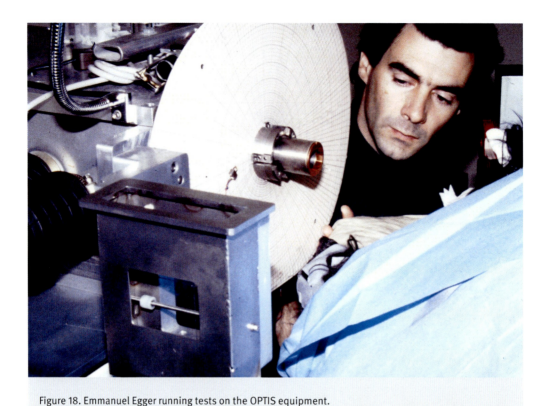

Figure 18. Emmanuel Egger running tests on the OPTIS equipment.

the other person would know what to do as well. In Orsay, she said, there were normally three people and they had to discuss everything all the time."

Providing an analysis of the growing bank of information about patient outcomes proved a massive task. Charles Perret had started developing software for the database and it was Egger's job to finish it. The idea was that when patients returned to Lausanne for a follow-up with the ophthalmologist three months to a year after their treatment ended, information from this meeting would be recorded on computers in Lausanne and then passed on to PSI for statistical evaluation.

"Since no personal computers were available at this time, the software was written in Fortran and running on the main SIN computer," says Egger. *"Data was entered by Professor Zografos's secretary, whose terminal was connected to the computer via a modem and telephone line. The data entry was very cumbersome, and it wasn't possible to correct an error after the 'enter' key had been pressed. Then, having done the first statistical analysis, we found that a lot of data were missing. By this time, PCs had become available, so I transferred the whole database to a PC and developed a new programme using a dedicated database software.*

"Over two years, my collaborators and I reviewed every single patient file and completed or corrected the information in the database. Then I began the detailed statistical analysis of the results of the treatment performed at PSI.

"We looked especially at those cases where the tumour continued to grow and needed a second treatment. I found that, in the early years of treating patients, we had about 10 per cent of local tumour control failures. I analysed those cases with the ophthalmologists in Lausanne, and we were able to identify the problems leading to the failures. Together with the ophthalmologists, we were able to find solutions and thus increase the local tumour control rate from 90 per cent to 98 per cent, which is really fantastic. This work resulted in the publication of several articles."

Egger regards this analysis as one of his biggest contributions during his time at PSI. It resulted in changes that reduced the risk of local recurrence, such as optimising the margins of the predefined target volume for irradiation.

"The results of Mr Egger's analysis were excellent, because it combined all our experience and helped guide us in localising the tumour, and we learned how to better designate the part of the eye that had to be irradiated," says Zografos.

Proton beam irradiation was also adapted for other ocular tumours such as choroidal haemangiomas, choroidal metastases, conjunctival melanomas and conjunctival carcinomas.

In an article written for an international proton radiotherapy workshop held at PSI in 1991, Egger, Zografos, Perret and Gailloud reported on the first seven years of proton beam irradiation of uveal melanomas. The progress made had been remarkable. Between March 1984 and the end of January 1991 a total of 970 patients had been treated – 876 of them for uveal melanoma. In only 3.5 per cent of patients did local tumour control fail, making a second treatment necessary. Most patients lost some visual acuity in the six months after treatment – but this was strongly associated with the size of the original tumour.

"There should be no doubt any more today that proton beam irradiation of uveal melanomas is a valuable treatment," they said. "The survival of the patients is comparable to the survival by other conservative treatments despite the fact that the tumours irradiated with a proton beam are much more voluminous than those treated by other methods.

"Unfortunately, up to now, just the very bad cases, i.e. tumors near the fovea [the central part of the retina] and/or the optic disc [the part of the retina above the optic nerve], very voluminous tumors, are treated with proton beams in Western Europe, so that it is not possible to say if there is an advantage for the functional prognosis using a proton beam in place of another conservative treatment."

The OPTIS project had demonstrated its value. Its progress had been closely watched around the rest of Europe, and by 1992 its success had inspired similar programmes in Belgium (Louvain-la-Neuve), the UK (Clatterbridge), France (Institut Curie, Paris, and Centre Antoine Lacassagne, Nice) and in countries outside Europe too – many using technologies developed at Villigen.

The project had also laid the firm foundations of expertise and technology that lie behind the sophisticated OPTIS2 programme at PSI today. The story of developments after 1992, including the new OPTIS2 facility, is told in Chapters 6–8. As with many PSI programmes, the success of the early years was based on collaboration and teamwork.

"This was a really excellent example of a joint project," says Zografos. *"When you talk of joint projects it often sounds like a political statement, but this really was a project where people worked together well. The idea came from PSI. They were looking for someone with experience of ocular oncology, and I was involved in the project from that moment. It has been very exciting to be involved."*

Chapter 4

1984–1992
From pions to protons

The success of the OPTIS programme in the years between 1983 and 1992 ran in parallel with another story at PSI, as the revolutionary techniques developed for the piotron evolved and then put PSI at the forefront of proton therapy innovation in the late 20th century.

OPTIS was demonstrating the potential of the proton for treating eye tumours – cancers which, because of their location, required comparatively low energy particles to get to their target and neutralise abnormal cells. But getting to tumours seated deeper within the human body was a different matter: it required particles travelling at much higher energies.

By the early 1980s, the piotron was showing the potential that pion particles had in this area. They could pass through tissue causing (in theory at least) relatively little damage and then release most of their energy at the target. What's more, the dynamic "spot scanning" technique developed by Eros Pedroni for the piotron was opening up a new means of delivering particles to their target with extremely high accuracy.

Thinking in three dimensions
It was this potential that enthralled Richard Greiner from the Department of Radiation Oncology at Inselspital, University of Bern, a consultant at the Department of Urology which referred patients with advanced bladder cancer to SIN for trial pion treatments between 1981 and 1984. On Carl von Essen's resignation as SIN's medical director, Greiner applied for the job. To him, it seemed a remarkable opportunity.

"Looking back now, I still think what a wonderful idea it was – absolutely unique at the time in Switzerland. It was a completely new type of treatment, with the patient moving, under computer control, so that the 60 converging beams could hit all parts of the cancer. The centres at Los Alamos and Vancouver were also researching pions on patients, but they were using static horizontal or vertical beams. This was a truly dynamic treatment.

"I think one of the reasons I was chosen as von Essen's successor was that I was Swiss – they wanted someone who could establish better contacts within Switzerland. It was difficult for Carl, coming in as an American. He came in like a lord, unknown to most people, but bringing in all these experiences and information, and this incredible innovation. I was very proud to follow him. It was like a vision of how radiotherapy should be planned and applied in the future."

Greiner took up the role in early 1985 following a four-week familiarisation period in 1984.

Carl von Essen had had limited success in engaging national interest in, or support for, the pion project. His efforts had often met with resistance from a Swiss establishment reluctant to change – a problem that plagued successive medical directors at PSI. Greiner remembers von Essen presenting information on the early pion therapy trials to Swiss radiation oncologists between 1981 and 1984. However,

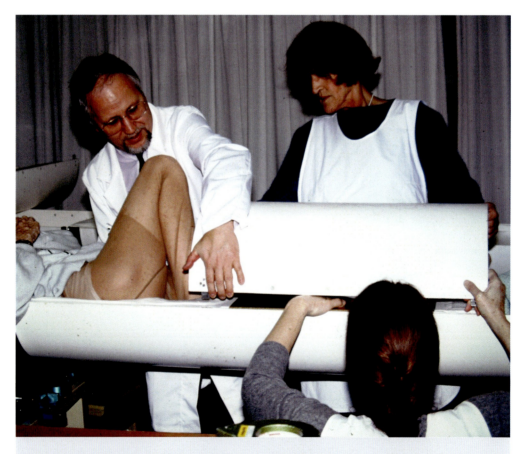

Figure 19. Richard Greiner helps prepare a patient for irradiation in the piotron.

they did not seem willing to understand the science of negative pi-mesons, and even less so the revolutionary idea of using dynamic targeting.

"The idea of a patient positioned inside the piotron was in itself very strange, and even more the patient's dynamic movement in three dimensions against the pion spot," says Greiner. *"No matter how fascinating and outstanding the medical radiation project at SIN may have been at that time, it did not really get known in the small family of Swiss radiation oncologists. There was a lack of interest and therefore a lack of knowledge, petty jealousy and concern about referral of cancer patients to such an industrial environment. The Swiss specialists were still thinking in one dimension."*

Compensating the power of the pion

This was challenge enough. But Greiner's enthusiasm took a blow when confronted with the disappointing results of the first trial on advanced bladder cancer patients. The local control rate of the tumours was quite high, but no one had predicted that pion treatment would bring changes to the connecting tissue around the bladder weeks or months after treatment had ended. The reasons for this needed to be understood. So, as trials on applications of the piotron continued, Greiner worked with physicists and radiobiologists to solve the conundrum.

Part of the problem was that the whole field of radiobiology – the study of the effect of ionising radiation on tissue – was still emerging. In the early 1980s, there was only limited understanding of the relative biological effectiveness (RBE) of particle beams. The RBE was not well documented for a single dose of any type of radiation, and certainly not for late-reacting connective tissue.

Greiner and his team systematically studied the documentation of the bladder cancer patients treated with pions between 1981 and 1985.

"We had to build understanding that radiotherapy could trigger late side effects, dependent on single dose, clearly stronger than early effects and clearly greater than initially anticipated," says Greiner. *"The only possible solution was to adjust dosing for each fraction of pion radiation in order to respect the tolerance of the surrounding healthy connective tissue. The new understanding gained from these early patients worked very well and was effective."*

Gudrun Goitein, who joined Greiner's team in 1988 and soon succeeded him, later argued that neutrons, freed upon the action of pions within the tumour, were causing a supplementary and wide-reaching dose in addition to the prescribed dose.[10] So the effect of a pion dose on late-reacting tissue was greater than initially anticipated.

Greiner, she says, intelligently adjusted doses accordingly. *"This worked very well,"* says Goitein. *"It just shows you how, despite having a really top-notch scientific committee with worldwide acknowledged experts, and despite radiobiological experiments with cells and animals, it is extremely difficult to translate experimental data into patient data and application on patients. It was the unusual physical characteristics of the pion particle that gave it its high RBE."*

It became clear, as experience with the piotron grew, that the particle was more advantageous for larger tumours.

Gudrun Goitein: *"We started with the intention that pions should be used to treat relatively radioresistant tumours like bladder cancer and glioblastoma. We came to learn the advantages of pions for treating sarcomas – which are very large tumours of the connective tissues. They can grow very slowly in areas where you don't feel them, so they are not easily resectable at time of diagnosis, and in adult patients they don't respond well to chemotherapy. With conventional radiation therapy you don't get the necessary dose to destroy them. But we treated some of these giant tumours pre-operatively and they became resectable within six to eight weeks."*

Convincing the outside world
Richard Greiner was certain that pion therapy held great advantages in treating large tumours, but getting patient referrals was not easy.

"We were a small country, and the piotron was partly backed by the Swiss Cancer League and the Swiss National Science Foundation. Every year I had to write papers to convince people that we should continue with pion treatment. But when the findings about the bladder cancer patients came out, it reflected on the whole project, and we had trouble getting referrals. I had to convince people to refer patients with big tumours, who otherwise would not find any help."

---

[10] Andreas Pritzker, *The Swiss Institute for Nuclear Research SIN* (Norderstedt: Books on Demand, 2014).

Greiner spoke widely about pion therapy and the revolutionary techniques surrounding it, such as dynamic scanning and computerised treatment planning based on CT scans. As he did so, he began to understand the challenges of being at the cutting edge of clinical change.

"During my time at the institute, we had excellent results and excellent connections. I went to international meetings with all the people pioneering high energy treatments – from Berkeley, Los Alamos and Boston. It was a very good time for me, the best time in my academic career.

"After I left, I had several positions, including chief of radiation oncology in Bern, and I took over so many things from my work at SIN. But it was not easy, because we had been pioneers, and now I had to fight for some things I took for granted. Ten years after SIN had begun planning treatment using computed tomographs, I still had to fight for the same thing outside that environment. So there was this gap of ten years between the institute and everywhere else."

The pion patient experience

All this – the pioneering nature of the treatments, the unusual scientific setting, the mixed results from early trials – meant that patients referred to the institute for pion therapy needed special support. Richard Greiner was all too aware that, as well as being extremely ill, most patients would also be extremely worried.

"The radiation project at SIN was part of a scientific and experimental area – a quite alien environment even for radiation oncology physicians, who are used to working in a clinical environment, consulting with other clinics, discussing and getting advice from other colleagues in difficult situations. All those things were lacking at SIN, and it needed habituation.

"And that lack of clinical environment was even more striking to patients suffering from cancer. This made the consultation with patients before treatment very important. The disease is not the only fear. The strange industrial environment was frightening and all the technical equipment created questions which had to be answered. The better informed patients were, the better they tolerated their time at SIN."

But the extent of their illness could be daunting.

"Tumours were often very far advanced – indeed, that was the very indication for sending patients to PSI for pion treatment. This presented a huge challenge for planning, preparation and positioning inside the piotron – especially for the dynamic three-dimensional movement. I remember a Scandinavian patient suffering from a sarcoma as big as three fists growing into the back of the head. It was very difficult to find the correct position for his head inside the piotron and even more challenging to find the right positioning for a special spot scanning technique."

There were emotional as well as intellectual challenges. At times, the direness of an individual's situation sent Greiner to the engineers and physicists to see if they could find ways around treatment problems. Eros Pedroni, at that point working outside the medical facility, was one of those he called on.

Pedroni remembers Greiner contacting him in 1985 to discuss a possible solution for a young man with cancer. The case affected Pedroni deeply.

"I felt sad and angry seeing the life of a young cancer patient being destroyed by a blind injustice from 'mother nature'," says Pedroni. *"The experience remained in my mind."*

That sadness and anger was, in the end, deeply influential. It planted a new seed in Pedroni's mind that he wanted to return to the medical division and build on the work he had begun. Two years later, he would do just that, and embark on work which would result in the first spot scanning system on a gantry in the world.

A shifting emphasis
Throughout the 1980s, interest in the piotron project grew. Most of the patients being referred weren't Swiss, but arrived from Italy, Spain, France, Austria and Germany. Richard Greiner recalls that some were celebrities and actors with serious tumours who wanted to keep their illness quiet.

"Sometimes helicopters came with patients – it was very strange, these aircraft arriving in this industrial area. We were often asked what was going on. Some people were coming for a first consultation to see if pion therapy would help them. Sometimes we would get calls from foreign ministry departments asking about exactly what could be done with this new treatment. There was a list of actors being referred to us by doctors who said they had run out of options."

There was growing interest too from the great and the good in medical research.

Richard Greiner: *"There was a young French-speaking woman with a two-litre, unresectable sarcoma growing around both kidneys and the small bowel. We agreed with her surgeon that the treatment aim at SIN was to get the tumour operable. We achieved that, and it was the biggest tumour volume ever successfully treated at SIN. During her treatment, we were visited by Professor Herman Suit from Boston – one of the most respected names in radiation oncology. He was very surprised that we had made this tumour treatable.*

"At that time radiation oncologists could not yet use CTs and computers for conforming the tumour and planning the distribution of the applied dose. And medical physicists were certainly not yet accustomed to a dynamic application of the beam. These possibilities we were demonstrating were unique."

The star of the show, it turned out, was not the pion particle but the treatment planning systems and dynamic scanning techniques that had been developed to deliver it. They allowed a radiation dose to be "conformed" to the exact shape of the tumour in three dimensions, sparing surrounding structures, in a way that had never been done before. And the scientists, clinicians and administrators at SIN were as aware of this as anyone. It was Hans Blattmann who provided the main impetus for developing these innovative techniques further at the institute, but not with pions.

"We were treating some patients at the piotron, but not many, and protons seemed to me a much more practical and safe way to go," says Blattmann. *"At the beginning, I was not sure, because there were other institutions that had already worked with protons, and we did not want to copy something from another institution but to build something new. In research, you don't just want to be copying someone else. We wanted to go a step further.*

"So what we did was very innovative, and the whole thing was developing in a way we could not have predicted in the beginning. One of the important elements we had was the fast scanning with the beam – Eros Pedroni did most of the thinking and programming on that. The beam could move fast and exactly deposit the maximum dose at points where we predicted, where we wanted the dose distribution in a three-dimensional volume. Before that, it was mostly only two-dimensional areas which could be irradiated, using collimators."

In the 1970s, Blattmann had spent a year working at the Berkeley Laboratory, which had been treating patients with protons since the mid-1950s. He had also visited Stanford, which had developed the medical pion generator that inspired the piotron at SIN.

"In Berkeley, we had already experimented with active scanning, but with low energy only, and we could not actually irradiate anything larger than mice. So the next step was to irradiate large volumes three-dimensionally using active scanning, and this had now been achieved at PSI with the pion project."

Blattmann discussed the next steps with PSI's director, Jean-Pierre Blaser, and got his support for researching active spot scanning further, this time with protons.

"I had many discussions with Professor Blaser," says Blattmann. *"He was always interested in our developments, and provided very positive input into what we should do and what we should do differently. Naturally, he had a lot of priorities outside our programme, but from the very first he was always interested in our progress.*

"Our funding came partly from the Swiss National Science Foundation, and when we had to apply for money it was always a help to have a famous Swiss person like Professor Blaser behind us. It was for quite some time important to find money from different sources, and that was also one of the jobs that had to be done before we could do experiments."

So in 1987, Hans Blattmann invited Eros Pedroni to return to the medical division to work on the development of active scanning techniques with protons. The terminology was now beginning to change. For the new project, active spot scanning with protons also became known as "pencil beam scanning", to reflect the narrow and unerringly straight streams of particles, roughly the width of a pencil, that would be delivered.

Pedroni was happy to accept the invitation. Like Blattmann, he was excited at the potential of using protons – particles that had a lower LET than pions but were extremely precise.

"Pions had too low a dose rate and poor dose localisation compared to protons and other ions," he says. *"In fact, the interest in proton therapy started immediately*

after the commissioning of the piotron. It became obvious that in terms of precision of dose conformity, proton therapy would be superior to pion therapy."

But Pedroni wanted to lead technological developments.

"I told Hans Blattmann that if I came back I would want to have some freedom in shaping the pencil beam scanning project. I explained to him that I wasn't a young student any more, so I wanted to have a little more responsibility, and to be able to define my playground. Science always has a lot to do with being free to explore new possibilities and their limits."

New institute, new ambitions
The establishment of a new project for developing pencil beam proton therapy for deep-seated tumours, alongside the piotron and OPTIS projects, coincided with significant institutional changes which served the initiatives well.

SIN had been founded in 1968 by the Swiss Federal Institute of Technology (ETH) in Zürich – a public technical university set up by the Swiss government to educate engineers and scientists. During the second half of the 1980s, a project team appointed by the ETH Board investigated the possibility of merging SIN with the scientific institute on the other side of the river Aare, the Swiss Federal Institute for Reactor Research (EIR). As the two neighbouring research institutes had evolved, they had developed overlaps in their studies of neutrons and protons, nuclear medicine and nuclear fusion technology.

The decision to amalgamate the two was made in 1986, and on 1 January 1988 they became one new centre of excellence in science: the Paul Scherrer Institute (PSI), named after the leading Swiss physicist who had championed research into high energy particles at ETH in the 1930s. There was to be a new emphasis at PSI: on the practical applications of physics research. This fitted well with the philosophy of Jean-Pierre Blaser, now appointed the first PSI director.

Among those on the ETH project team researching and implementing the merger was Martin Jermann, who had been manager of research and development at EIR, and joined the new institute as head of staff for strategic scientific development. Before the merger, he had been aware of proton therapy, and knew something of the piotron project because of his contact with its designer Georg Vecsey through work on fusion energy at SIN and EIR.

When Jermann started his new role, particle therapy was a very small part of PSI's activities. The budget for research and development at the new institute had to cover basic science, energy, materials science and solid state physics, as well as biology and medical applications. But early in his tenure Jermann became excited about the potential of particle therapy research – and not just because it might help many incurably ill people. He also saw that it could play a valuable part in positioning the institute. Applied research and socially relevant projects gave it direction and identity to outsiders. This became very clear in 1989, when the new PSI was visited by a parliamentary financial commission of the Swiss government.

Martin Jermann: *"We were starting building a big new project, the spallation neutron source, based on the large proton accelerator facility. It would be the first of its type in the world. The main goal of the visit was to review the project plan and to approve a credit of around 35 million Swiss francs to help fund the project.*

"I remember that one member of the commission was totally against this project because it had to do with neutrons. She was a very important and interesting person – one of the first women in the Swiss Parliament. She had heard of neutron bombs, and was worried about atomic energy, and she said she wouldn't support the neutron source project.

"Before the commission made its final decision, it visited the construction site where the neutron source was due to be built. So we went over the river to the site, then walked back over the bridge towards the meeting room. As I was walking with her, we passed the piotron facility. I explained to her that we treat cancers with particles here, and asked her if she would like a quick look before going back to the meeting room. She agreed.

"As we went into the treatment room, they were just preparing a 19-year-old man for treatment with pions. He had a glioblastoma. Hans Blattmann was there, and I introduced him and he explained the facility, and then Richard Greiner also came in, so he demonstrated the clinical part. It was very emotional, because it could be seen that this young patient had already been treated with chemotherapy and – as Richard explained – there was a very low probability of cure. But, with the new technology, using pions, we hoped to get better results. And so he was treated with the piotron.

"She was very impressed, and on the way back to the meeting, she changed her mind about the spallation neutron source. It was because the source of the neu-

trons was the proton, which also produced the pions used to treat cancer patients. So she supported the neutron source project, including an upgrade of the proton accelerator.

"This experience triggered me to make this area of work a priority in the overall research programme at PSI in the coming years."

Planning for a new system of proton therapy
According to Eros Pedroni, there was a definite change of outlook when PSI was founded, as he began to work intensively on pencil beam scanning.

"It had become clear that basic physics alone would not be enough to maintain the old institutes, and that was behind the formation of PSI. PSI was moving in a new direction. Yes, basic science, but connected with applied physics as well.

"So I thought that after my experience with pion therapy, there was the possibility of doing something practical with protons. The contact that Richard Greiner had made earlier was important, because it helped me understand that there was merit not only in doing something for science, but for society.

"There were physicists who were much better than I am, but I always thought that if I could do something related to basic physics that was also of use to society, then this would help science too, because politicians and the public are interested in the applications."

Something else was changing at PSI. SIN had not had the right energy range of proton beam to treat deep-seated tumours with protons. The 72 MeV protons being produced by the injector cyclotron for eye tumours were not powerful enough to get deep into tissue, and the 590 MeV protons from the main ring cyclotron had far too high an energy.

The situation changed when a new facility for physics experiments was built in a new hall (named the NA Hall) near the piotron building, to conduct experiments on polarised proton-proton scattering. The piotron's beam line was extended into the hall, and then degraded from 590 MeV to variable energies, using a degrader unit. It was this degraded beam line, of 100–200 MeV, that would be used for the proton therapy experiments.

"The physicist Manfred Daum was in charge of the realisation of the NA complex, and he was very supportive, permitting our last-minute intrusion into his ongoing project," says Eros Pedroni. *"It meant that we could start looking at proton therapy for deep-seated tumours at PSI."*

So by 1988 the development of dynamic pencil beam scanning with protons had become more technically feasible, and within a more favourable environment. For Pedroni and Blattmann, this was an opportunity to do something genuinely innovative with protons – to take a different tack than American innovation, which had focused on a proton delivery technique known as "passive scattering" (see "Passive scattering vs active scanning" box).

What Pedroni and Blattmann envisaged was completely different: a pencil beam scanning system as part of a mechanical apparatus known as a "gantry", which would deliver multiple fields of beams without requiring patient repositioning. This, they believed, would increase accuracy, reduce exposure of healthy tissues to radiation and reduce treatment times compared with scattering. As with the piotron, they wanted a dynamic system that worked in three dimensions, moving the beam within the tumour. But whereas the piotron moved the patient in relation to the 60 beams, the gantry would move a scanning beam around the patient.

Eros Pedroni: *"At that time, the first hospital-based proton therapy facility in the world was being realised at Loma Linda in California. The facility included three rotating gantries – the first proton gantries in the world. The beam delivery technique used in Boston was based on the scattering foil technique developed at the Harvard cyclotron in Boston, the leading facility at that time. A possible alternative for PSI was the development of a compact proton gantry based on proton pencil beam scanning."*

Passive scattering vs active scanning
PSI's innovation in proton therapy hinges on the fact that, at a time when most of the world was seeing proton "scattering" as the way ahead in treatment delivery, PSI took forward a new technique called "spot scanning" (also known as "pencil beam scanning", or PBS). The first experiments on proton spot scanning took place in Chiba, Japan, in the late 1970s, but the technology was not further developed. In the late 1980s and early 1990s, proton therapy projects in Belgium, England, France, Japan, Russia, Sweden and the United States all went in a different

direction. It was called "beam widening", of which the most common variety was "passive scattering".

In the scattering technique, the proton beam is widened so that it covers the entire volume of a tumour. This is achieved by directing it at a slab of material (usually a thin metal foil) where the protons are randomly deflected from their original direction, considerably enlarging the cross section of the beam. Then a collimator cuts the widened beam profile to the shape of the tumour. A "compensator" can also be used to better adjust the dose to the tumour volume. Normally, collimators and compensators must be individually manufactured for each irradiation field.

With the scanning technique, in contrast, the proton beam is kept unscattered, and the Bragg peak only irradiates one part of the tumour volume at a time. The beam is deflected by means of "scanner" magnets so that the next partial volume can be irradiated. The process is repeated until the entire tumour volume has been scanned.

For both methods, the energy and thus the range of the protons must be controlled. In the case of passive scattering, this is usually done with a "range modulator wheel" – a rotating disc of varying thickness. Pencil beam scanning requires a set of different energies to cover the full range of the tumour. In the case of a cyclotron, an accelerator that can deliver just one fixed energy, the energy change is achieved with a "degrader", a set of wedges with varying overlap inserted into the beam.

The advantage of scanning is that the dose can be better conformed to the target volume, especially if it is irregular. No tailored hardware is needed for each patient, as the scan sequence is controlled by a computer, and any dose distribution can be generated. It was this feature that made the development of the Intensity Modulated Proton Therapy (IMPT) technique possible.

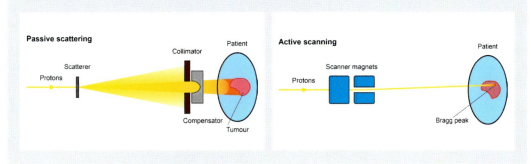

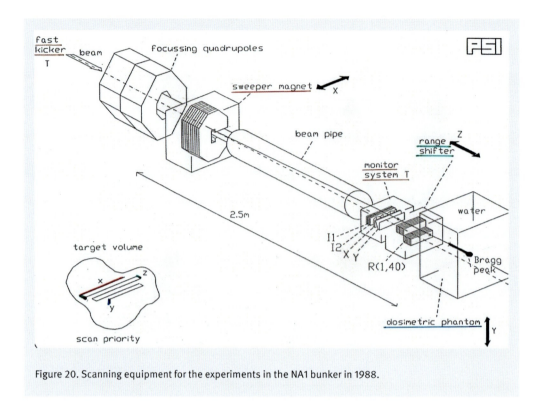

Figure 20. Scanning equipment for the experiments in the NA1 bunker in 1988.

Experiments led by Pedroni and Blattmann on what was known as the NA1 beam line began in a provisional bunker within the experimental area. Using magnets, the beam line and its range were shifted in two dimensions, with the target – a patient phantom – being lifted and lowered in the third dimension.

Each "spot" was irradiated for milliseconds before the beam was shifted to the next spot, with the amount of dose per spot measured with two intensity monitors. The beam position was also monitored. As expected, the sharpness of the dose of the proton pencil beam was significantly superior to the pion beams of the piotron.

As a feasibility demonstration, a radiation-sensitive film was irradiated with the scanning system, writing the numbers "1990". The first version of dynamic pencil beam scanning with protons had been successfully put into operation. To mark the success of the experiments, the image was used for a seasonal greetings card distributed internally at PSI as the year ended.

"I well remember the astonishment when these proton-written numbers were shown at the next meeting of Swiss radiation oncologists," says Richard Greiner.

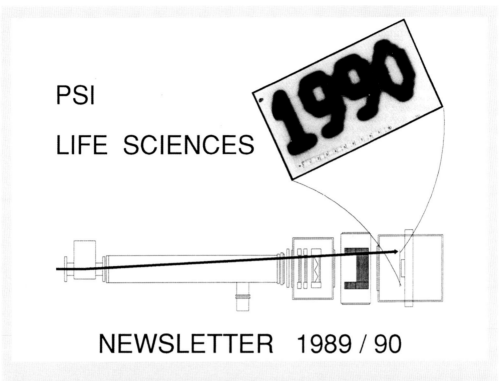

Figure 21. The first scanning result – "1990" written on radiation-sensitive film with a proton beam – was used on the front page of the Life Sciences Department newsletter.

At the same time, one of Pedroni and Blattmann's collaborators, Dölf Coray, recalculated a pion treatment case using protons. It demonstrated how attractive the dynamic methods developed for the pion therapy project were when applied with a sharper beam of protons.

Pedroni and Blattmann worked up their proposals to deliver proton scanning via a gantry.

"In 1989, Hans Blattmann came into my office and presented to me the proposals for the financing of the gantry project," says Martin Jermann. *"They had already made detailed designs, but now it was important for them to get financial backing for the manufacturing and construction of the gantry. They needed an industrial company to build the mechanical structure of this gantry and also some equipment like special magnets. So they needed up to two million Swiss francs, mainly for the contract with the engineering company. I talked with Hans about the proposal and said: 'Leave it with me.'*

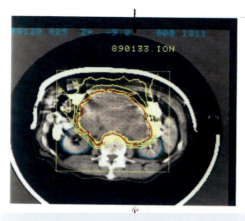 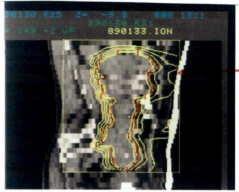

Figure 22. Coray's recalculation of a pion treatment if protons were used – one of the very first scanning proton therapy plans made, using the piotron's treatment planning system. The tumour, delineated in red as the target volume by a medical doctor, is visible in both CT scans – from above (axial) on the left, and from the side (coronal) on the right. Organs at risk (the kidneys) are delineated in blue.

"At that time, the decision was not mine: it was up to the director, Jean-Pierre Blaser, and his deputy, Wilfred Hirt. I told this to Hans. I had a meeting with Wilfred Hirt and we looked through the proposals, and he told me that budgets were tight, and we had recently made major commitments to a cryogenic facility for fusion technology research and the spallation source. These were high-priority projects for PSI. But he said: 'If you are convinced that this is also a good project for the future of PSI, you can decide.' And I remember an hour later I signed the procurement application and sent it to the purchasing department."

In September 1989, PSI's growing status in the fast-evolving world of particle therapy was marked when the institute hosted the annual meeting of PTCOG – the Particle Therapy Co-Operative Group. It was the first time a PTCOG meeting had been held outside North America. PTCOG had been established by a group of scientists in 1985 to develop understanding of the clinical applications of particle therapy and encourage the development of hospital-based particle therapy facilities. Hans Blattmann had already given a presentation to the group in 1986, reporting on the experience of conformation radiotherapy delivery with pions, and its possible relevance to proton therapy.

In its early days, PTCOG was a small community dominated by American scientists. So this first meeting in Europe was significant. With all other proton therapy initiatives having decided that scattering approaches were the way forward, there was curiosity about PSI's alternative scanning techniques.

The 116 registrants at the meeting at Villigen on 18 September 1989 were welcomed by Jean-Pierre Blaser and the PTCOG chair, James M. Slater, the man who had initiated the first hospital-based proton therapy centre in the world at Loma Linda University, California. First on the programme was Eros Pedroni, who described his 200 MeV proton therapy scanning project. Later there was a tour of PSI, with visits to the piotron, the OPTIS facility and the proton scanning experiment.

Figure 23. Eros Pedroni.

It was decided that the next meeting, in 1990, would be held at Loma Linda, and should coincide with the commissioning of the new proton medical facility there. "It has been suggested that we hold a workshop on gantry design (including beam scanning issues) in conjunction with the next meeting," say the minutes.

A change of medical director

At the end of 1989, Richard Greiner left PSI after nearly five years. He returned to the Inselspital at the University of Bern, and remains proud of what was achieved while he was at PSI. There had been challenges, many financial. Despite the clear advantages of pion treatment for some tumours, it had been a continual struggle to get reimbursement for treatments from the Swiss authorities. And there were technical obstacles too.

"The strongest challenge for treatment with pion beams became our ability to deliver on our promise to the patient that we would apply the appropriate pion dose to their tumour within the optimal time period," says Greiner. *"The cyclotron needed annual checks and repairs, and needed to be shut down for maintenance between December and April every year. So there was no proton production, and therefore no pions. Sometimes there were also problems because of beam instability, which made interruptions of the sessions necessary."*

But Greiner's tenure represented, as he puts it, an important "hinge" as PSI shifted emphasis from using pions to treat deep-seated tumours to using protons instead. The use of protons in the OPTIS programme continued all the while in parallel.

"I was a manager. I could not be a pioneer. But I was the first Swiss physician to use the techniques of CT planning and conforming, and I was very proud of that and my position at PSI. I am really part of the history of Swiss radiation oncology because I am part of many eras in radiation therapy."

His successor was to oversee the transformation of the experimental pencil beam scanning programme into a treatment facility successfully treating scores of cancer patients every year.

Gudrun Goitein (known as Gudrun Munkel before her marriage to Michael Goitein) was a German diagnostic radiologist who had trained in nuclear medicine and became fascinated with PSI's piotron work during an extended visit while working as a radiotherapist in Zürich. On her return to radiation oncology in Germany, she was dissatisfied with conventional approaches. "I knew it wasn't the way I wanted to go," she says. When she heard there was a vacancy on the piotron project, she didn't hesitate to apply. She got the job, and a few months later was appointed Greiner's successor.

"For me, the most important outcome, the lesson from the pion project, was that it's not so much the radiobiological efficiency of the particle that is important," says Goitein. *"It is how well we can conform a high-dose region to an irregularly shaped mass in the body, at the same time protecting the nearby healthy tissues and cells as much as possible."*

Goitein backed Blattmann and Pedroni's idea of pursuing that lesson with protons. At the same time, she saw that the piotron's increasingly regular failings were jeopardising effective treatment.

Technical issues
It was Goitein who put an end to the pion project. *"The technology of the piotron was genius but very complicated. The system used to break down at the most inconvenient moments, so we often had to work up to midnight because it had taken hours to get the machine repaired. When you start a patient treatment but then the system breaks down after 10 treatments in a course of 25, it's irresponsible to continue."*

In 1989, the computerised treatment planning system went down when a technician accidentally wiped all copies of the software (stored on "layered" hard disks) by inserting them into a faulty driver. Eros Pedroni was called on to help with the emergency. Fortunately, he found in his office an old backup of the original software on tape and some code printouts. Four or five people were engaged to rewrite the code based on the tape and paper printouts, then the entire system had to be exhaustively retested. In all, the piotron was shut down for two months.

There were other interruptions. Between March 1990 and March 1991 all pion treatments were stopped while the PSI accelerator was upgraded and the experimental area was reconstructed. That year, and the other periods when the piotron was offline, provided valuable learning periods for Goitein, who, in the absence of patients, took the opportunity to find out all she could about the technology, physics and biology of particle therapy.

"For many years, I shared an office with Hans Blattmann – and you couldn't have a better teacher. I got my understanding of physics from Eros Pedroni. He said: 'What would you like to know?' He told me what was possible and, most importantly, explained why it was possible. I was reading all the time.

"After the pion treatments stopped, there was a period where we had only eye patients and I went to visit Boston and other facilities that were planning proton facilities. I stayed there for months to learn about their treatment planning and other aspects of their work in order to define our own project.

"I was responsible, together with Hans Blattmann and Eros Pedroni, for creating a project that would treat patients effectively. So it was important that I learned what was possible, what I wanted, and whether what I wanted was possible. I also learned what I did not want.

"I had been captivated with the idea of the moving beam and better conformation of the high dose – so I knew I didn't want the technology that was established at the Massachusetts General Hospital in Boston. They used scattering, sending a proton beam through some material to spread it out, and then sending it to the body. Then, because they didn't want to shower the entire body with the beam, they looked at the shape of the tumour from the direction of the beam – the beam eye – and put a lucite or wax collimator into the beam to block the parts that were not needed. They had to create hardware for every different angle they wanted to bring the beam into the body. That requires technical personnel and takes time. And that's the thing I didn't want."

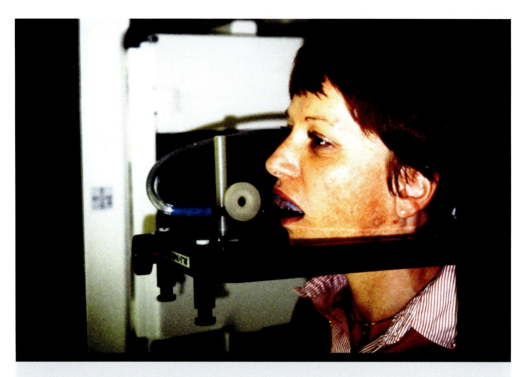

Figure 24. Gudrun Goitein (Munkel) was keen to try out all the equipment involved in patient treatment. Here she tries a bite block, used for patient positioning during proton therapy.

The proton project takes shape

Using the NA1 beam line to perform pencil beam scanning experiments had only been a temporary measure. The proton therapy research needed its own dedicated horizontal beam line, and permission to build this was granted in 1991. The line, named NA3, was designed by Eros Pedroni. It was to incorporate the first two requirements for a scanning system: different energies of the beam could be selected, and it was to have a "fast kicker" magnet system for ultra-fast switching on and off of the beam.

Eros Pedroni: *"The NA3 beam line was financed by dividing money from other projects of the newly constituted Life Sciences Department, headed by Edmund Löpfe. But these were times of uncertainty due to the scientific redirection of PSI after the fusion of SIN and EIR, and there were competing priorities. Having some experience in beam optics from my PhD student days, I was able to find a solution for the layout and beam optics of the NA3 beam line and gantry. The gantry beam optics were further reviewed and optimised by Harald Enge, an emeritus professor at MIT in the United States, known as the 'pope' of spectrometer design in the nuclear physics era.*

"Fortunately, the engineering divisions were very supportive and we could borrow many components from reserve material stored at PSI. The NA3 beam line was 'the foot in the door' for what would follow. Without improvisation and personal connections within PSI we would have had nothing to show at the time of the arrival of the new directors succeeding Jean-Pierre Blaser. This was, in my opinion, the most critical phase for the long-term survival of the medical division at PSI."

There was a sense of urgency about the new proton project. PSI was not the only institution forging ahead in this field. As Hans Blattmann wrote in an article for *Radiation and Environmental Biophysics*, "charged particle patient treatment . . . is entering a new decade of growth and development".[11]

By 1991, patients were being treated with protons in the USA (Berkeley, Harvard, Loma Linda), Sweden (Uppsala), USSR (Dubna, Moscow, Leningrad), Japan (Chiba, Tsukuba), England (Clatterbridge), France (Orsay, Nice) and Belgium (Louvain-la-Neuve).

As Blattmann pointed out in his article, all bar one of these projects had been based on machines built for physics research. This had influenced the application techniques used in particle therapy, since all the beam lines were fixed and horizontal. This was now beginning to change. Proton therapy clinical facilities were being planned within research institutes and hospitals using beam lines optimised for medical use. Different delivery techniques were also under investigation – among them, the scattering techniques developed at Harvard and used in Boston, and the conformational spot scanning being developed at PSI.

By 1991, the Loma Linda hospital-based proton therapy facility, with its 360-degree rotating gantries, had been launched.

"The time window of opportunity for a physics lab like PSI to substantially contribute to this field was slowly closing down," says Eros Pedroni. *"It was the right and probably last moment for starting a new project switching from pion to proton beams and using dynamic scanning as the technology."*

---

[11] H. Blattmann, 'Beam delivery systems for charged particles', *Radiation and Environmental Biophysics*, 31 (1992), 219–231.

But so adventurous were the plans at PSI that there were no direct competitors. It wasn't just the use of spot scanning that distinguished the PSI proton project from Loma Linda. While the Loma Linda gantries had a diameter of more than 10 metres, the PSI "compact" gantry would have a diameter of just four metres – potentially making it a far more practicable proposition for hospital installation. No one else in the world was attempting pencil beam scanning with protons using a compact rotating gantry – and computerised systems that planned treatment in three dimensions.

This was to become the model for nearly every proton therapy centre in operation around the world today.

In 1991, scientists from all over the world – including representatives from the proton projects in the Soviet Union, Sweden and the USA – came to PSI for a "Proton Radiotherapy Workshop" where Hans Blattmann, Eros Pedroni, Emmanuel Egger, Gudrun Goitein and others reported on the status of proton therapy at PSI and plans for the future.

OPTIS, they reported, had treated 198 patients in 1990, bringing the total treated there to 950. A copy of the famous PSI OPTIS chair had been installed at Linda Loma.

There was news of the progress made regarding health insurance and pion treatment, with the Swiss Association for Extended Health Insurance having agreed to reimburse for bladder and prostate cancer, bone and soft-tissue sarcoma and large gynaecological cancers.

And as for the experimental proton therapy project, it was envisaged that the first proton beam would go through the new horizontal beam line at the beginning of 1992, with the first patient treatments (without gantry) planned for later in the year. Based on Pedroni's concepts, the mechanical gantry design was currently being worked on by an external engineering company, funded by the Swiss Cancer League. The aim was to have a new treatment room prepared by the beginning of 1993, when gantry installation was to begin.

Blattmann, Pedroni and Goitein set out their case for a proton therapy gantry using a dynamic scanned beam.

They wrote: *"Even at a time when the first clinical proton therapy facility is being commissioned at Loma Linda, California, there are essential, basic questions which can best be investigated at a large research institute like PSI, where expertise in radiotherapy, medical radiobiology and accelerator physics as well as technical experience and skills are concentrated.*

"PSI has a strategy for its contribution to research of cancer management which combines efforts on the local, the loco-regional and the systemic level of the disease . . . The goal of the proton project is the development of a treatment setup with the highest possible flexibility to evaluate the parameters necessary or desirable for an optimal clinical proton therapy facility. The experience gained by actually treating patients . . . will yield crucial information to design a hospital based proton therapy facility . . .

"Apart from these strategic considerations, many practical problems in connection with dynamic radiotherapy have to be solved. These are connected with safety of the application of dose, particularly relevant for a dynamic application technique as the spot scan, as well as the positioning and fixation of the patient . . . and last but not least the treatment planning which has to be performed in a reasonable time on the basis of multiple CT slices and other diagnostic information."

The agenda was set. The new horizontal proton beam line, dedicated to the pencil beam project, came into operation in summer 1992. And, as construction work began on the treatment room and gantry, the equipment from the NA1 bunker was used to conduct new research on the NA3 beam line. The aim of treating the first patients by the end of the year was already looking optimistic.

The end of the piotron

By 1992, just over 500 patients had been treated under the pion project. But the proton therapy project was now overshadowing it, taking the best of its remarkable innovations forward in a new direction. "Looking back, pions have not turned out to be the magic pill one had wished for," noted the PSI internal newsletter Spektrum.[12] After a final disastrous failure of the cooling system for the superconducting magnets, the piotron was closed down. The pion projects at Los Alamos and TRIUMF also ended.

---

[12] As reported in *The Swiss Institute for Nuclear Research SIN by Andreas Pritzker* (Norderstedt: Books on Demand, 2014).

But the piotron did provide a springboard for the remarkable innovation that occurred in proton therapy at PSI over the next two decades.

"Pion therapy know-how was the background for all subsequent developments," says Eros Pedroni. *"It was a fantastic interdisciplinary experience involving medicine, physics, radiobiology and engineering. But it was not the right particle. The production of pions as secondary particle beams produced too much radiation, and dose localisation was poor compared to protons. With the development of high energy synchrotrons, radiotherapy with heavier ions like carbon was becoming possible. These ions behave similarly to protons but are high LET particles, and today they present the best option for investigating the use of high LET radiation in medicine.*

"But though it didn't produce, pion therapy was historically a breakthrough. It's an example that shows how in science, one should go and explore the different possibilities and take the risk of not having success. It was an excellent project."

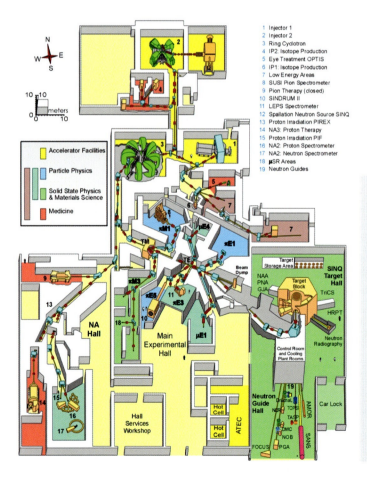

Figure 25. Diagram of the accelerator facilities at PSI in the 1990s, after the piotron had been shut down. The medical facilities can be seen in red, with the NA3 proton therapy area and gantry at the bottom left corner, and the disused piotron above it. The OPTIS facility is at the top right of the main hall, with the Injector 1 cyclotron which provided its protons just above (in turquoise). PSI's main ring cyclotron is to its left, in green.

Chapter 5

1993–1996
A revolution in cancer patient treatment

In a key paper, written in 1993 and published in Medical Physics in January 1995, Eros Pedroni, Hans Blattmann and Gudrun Goitein announced to the world the "conceptual design and practical realization" of the new proton therapy facility being built at PSI.[13] The project, which had begun with the building of the NA3 beam line in 1991, was of "unique interest" because it would perform the new pencil beam scanning technique via a compact rotating proton gantry.

"With these new technical developments at PSI we hope to be able to give a further significant impact to the success of proton therapy in the future ... With this project we also expect to acquire the experience necessary for the design of the next generation of proton facilities."

The aspiration was to be more than fulfilled. PSI's innovative – and controversial – project to combine proton pencil beam scanning with a gantry-based delivery system was set to become the model for proton therapy centres throughout the world. The road leading to global recognition would be fraught with daunting technical, financial, institutional and human challenges. However, PSI had deliberately not chosen the easy course.

As Hans Blattmann says, innovation was the key. There was no point in a science research institution like PSI simply copying what had been done in the United States.

"We wanted something new, to go a step further than they had in the US," he says. *"The fast scanning of the beam was the most important thing. The beam could*

---

[13] Along with their collaborators Reinhard Bacher, Terence Böhringer, Dölf Coray, Antony Lomax, Shixiong Lin, Stefan Scheib, Uwe Schneider and Alexander Tourovsky: E. Pedroni, R. Bacher, H. Blattmann, T. Böhringer, A. Coray, A. Lomax, S. Lin, G. Munkel, S. Scheib, U. Schneider, et al., 'The 200-MeV proton therapy project at the Paul Scherrer Institute: conceptual design and practical realization', *Medical Physics*, 22:1 (1995), 37–53. doi: 10.1118/1.597522. PMID: 7715569.

move very quickly and exactly deposit the maximal dose at specific points where we predicted we wanted the dose distribution, covering a three-dimensional volume. Before, using collimators, people had been irradiating two-dimensional areas. The new thing we started was three-dimensional point irradiations."

The key to successful radiotherapy was conforming a high dose throughout an irregularly shaped tumour, while at the same time protecting the nearby healthy tissue and cells as much as possible. PSI believed that the pencil beam scanning gantry system was the best way to achieve this treatment "optimisation". It needed pursuing, regardless of the risks.

The plan for the beam line and gantry

To optimise dose delivery, the gantry project effectively combined several new and advanced technologies – the proton pencil beam scanning systems pioneered at PSI, a new compact gantry design, advanced imaging technologies such as CT scans and the computerised treatment planning systems successfully developed for the piotron.

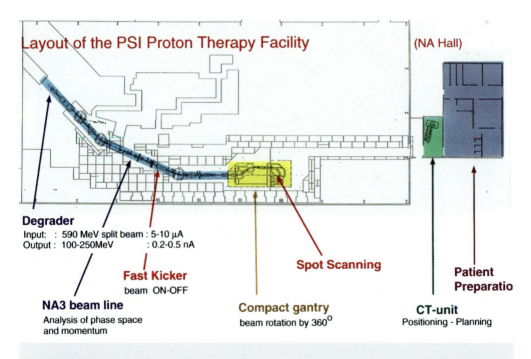

Figure 26. A contemporary diagram, prepared for overhead projection, showing the layout of the proton therapy facility in the NA Hall, with the new dedicated NA3 beam line, the patient preparation pavilion and the area for CT scanning.

The new NA3 beam line of 85–270 MeV would travel to the gantry within a "treatment facility" in the NA Hall. The energy of the beam could be adjusted by changing the arrangement of the degrader blocks and magnet settings on the beam line, meaning it could be set at a value selected for each patient and for each field of treatment delivered. The energy within a dose field could be modulated with a range shifter.

Outside the treatment facility would be a patient preparation area, or "medical pavilion", where patients could be prepared for treatment. Their positioning on the treatment couch would be verified before each fraction in a CT-scanning room beside the entrance to the treatment area. Anatomical landmarks recorded in treatment planning CT scans would be matched with new scans at each session.

The layout of the gantry had been finalised in 1991 and its mechanical structure was now to be fabricated by a Swiss engineering company. It was to be four metres in diameter and would weigh around 110 tons – much of the weight accounted for by the large magnets required to bend the proton beam around the structure to the patient. One of the magnets would be a "sweeper" magnet – which shifted the proton pencil beam from side to side horizontally. This formed one axis of the three-dimensional "scanning" movement of the proton beam. Movement on the second axis would be achieved by changing the proton range using the range shifters. The third axis of movement was created by moving the patient table in a direction perpendicular to the plane of beam sweeping. Together, these movements meant that any distribution of proton dose could be "painted" in three dimensions to any shape.

The gantry head would rotate around the patient, stopping at particular points to deliver proton doses from specified directions (dose fields), with the beam emerging from a "nozzle" directed at the patient. Every set-up movement and beam delivery had been programmed into a multi-field treatment plan and would be executed automatically under remote control. Collision shields around all moving elements would protect the patient's safety, and any detected error would stop the beam and all motors.

"The space in the NA Hall was almost completely occupied by equipment for physics experiments," says Eros Pedroni. *"To fit, the gantry could have a radius of only 2 metres, compared to Loma Linda's 5.5 metres. So we had to use many tricks to make the gantry very compact.*

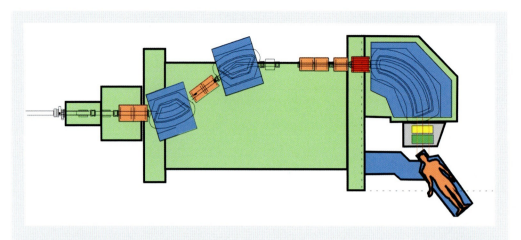

Figure 27. Diagram of Gantry 1. The magnets bending the beam line are in blue, with the sweeper magnet (which produced "scanning" movement) in red. On the right, the proton beam is bent 90 degrees by the largest magnet, then passes through a nozzle (in grey) containing detectors for proton dose and position (yellow) and a range shifter (green), before reaching the patient. When the gantry rotates, the patient table moves with the gantry, but a counter-rotation is applied to the table support so that the table remains horizontal at any gantry angle.

"One trick was to place the sweeper magnet before the last 90-degree bending magnet, which saved a couple of metres. Another was to mount the patient table on the front wheel of the gantry with an offset from the gantry axis, which saved another 1.1 metres. The support holding the patient table had to counter-rotate against the rotation of the gantry to maintain the couch horizontal at any time.

"With the addition of a further rotation axis mounted on the patient table it would be possible to rotate the patient couch also in the horizontal plane. This would allow a patient in the supine position to be irradiated from essentially any direction in space."

The realisation of spot scanning required not only a big technological effort to provide the hardware, but also the development of completely new treatment planning software. Pedroni, Blattmann and Goitein wrote in 1993: "Here we can rely heavily on our experience with dynamic scanning acquired in the past from the pion therapy program at PSI. The treatment planning system is now being completely re-written for the protons."

The need for safety as well as efficacy was paramount. Eros Pedroni says it was carefully built into plans from the very start. *"In designing the machine, we introduced the necessary hardware to ensure that the ongoing scan proceeded cor-*

rectly. For example, we would use two independent computers, each working with different but consistent prescriptions. This approach would avoid all common causes of failure and make the system inherently safer and, in the end, easier to commission."

However, for all the care being taken in the planning, there were still many unknowns.

"We should not forget that the 1980s and 1990s were a different time in terms of technology," says Gudrun Goitein. "There was so much to plan. How would we create the project? How strong should the beam be? How do we achieve this? How would we stably position patients? These were very important questions, because if you have a fast-moving beam, you're very sensitive to body motion. Particle beams have ballistic characteristics and behave differently according to what type of tissue they go through, and it is enormously technologically complex to make sure that particles consistently meet the target where you want them to.

"It was very important for me to understand all the issues involved, because in the end I was responsible for what happened. I'm the physician who says go ahead. I think the fact that I had these people around me – Hans Blattmann, Eros Pedroni, Dölf Coray – helped a lot. They were so profoundly engaged and knowledgeable, and because they had been involved in the pion and eye projects, they were already used to thinking in medical terms. They knew this was not like other physics experiments. I think this was the most important time for the project: this very thorough preparation. We knew what we wanted to do and why we wanted to do it."

It helped to know that the whole institution was behind the project. Jean-Pierre Blaser had retired as director of PSI in 1990. He was succeeded briefly by Anton Menth and Wilfred Hirt, and in 1992 Meinrad Eberle began a ten-year tenure. Without Eberle's strong backing for radiation medicine projects at PSI, the gantry project would never have got off the ground.

"He was very supportive," says Martin Jermann, who became Eberle's head of staff in 1993. "It was a good collaboration. When PSI put forward its proposals, there was a lot of scepticism from the outside world. It was too challenging and complicated, they said, and it was far better to keep with scattering techniques. It was my job to bring resources in, and I was always confronted with the question 'Why do you want to go in that direction? It costs a lot of money and will never be *necessary.' But we said: 'Let's do it. Let's see if there are advantages over passive scattering.'"*

Financial support also came in from the Swiss Cancer League and the Swiss National Science Foundation.

The spot scanning beam
The focused gantry beam had a width of 7–8 mm as it emerged for patient treatment. This compared with the pion beam width of 3 cm. The dose spots (the dose maximum at the Bragg peak of the beam) were applied in a grid pattern, with a grid spacing of typically 5 mm. The beam was switched on and off for each spot using the kicker magnet in the NA3 beam line – the length of the on sequence could be controlled within 50 microseconds for each spot dose. Both the intensity and position of the beam were measured during spot scanning with transmission- and position-sensitive ionisation chambers mounted in the nozzle (the point where the beam emerged).

Optimising the dose
During the second half of 1992, Pedroni, along with Shixiong Lin, Dölf Coray and Terence Böhringer, focused on treatment planning tests using the NA3 beam line. They examined how different parameters could interact to deliver precise doses: beam energy, range shifting, calibration of monitors, the amount of air the beam had to travel through before arriving at the patient. Working with them was PhD student Stefan Scheib, whose task it was to provide a new state-of-the-art computer code for proton pencil beam scanning, following the concepts developed in pion therapy.

The work involved modelling how proton beams might scatter when passing through air and tissue, affecting their dose peak (the point at which most energy is lost). Those models then had to be tested in experiments. Water was often used as a substitute for human tissue.

The work produced the first "optimised" treatment plans – computerised calculations of the right dose of protons for every one of tens of thousands of spots scanned within the three dimensions of a tumour. This experimental work, presented in Scheib's thesis, showed that the system was capable of "painting" the dose with high precision in three dimensions and that, in principle, patients could be treated in a horizontal set-up.

Figure 28. Tony Lomax, shortly after his arrival at PSI.

These innovations were fascinating to Tony Lomax, a young postdoctoral researcher from Manchester in the UK, who joined the project team in 1992 to work with Pedroni, Scheib and Böhringer on treatment planning. Lomax had worked as a medical physicist in UK hospitals and had focused on imaging in his PhD. He found the work in the UK interesting but routine. When his wife got a job at the University Hospital of Zürich, he moved with her to Switzerland for a new challenge.

"What's interesting is that when I moved here in 1992, there was a very small number of facilities doing proton therapy around the world, and all but one or two were based in physics research institutes like this one," says Lomax. *"It was very exciting coming to PSI, and a complete change.*

"For my first four years, there wasn't any clinical work at all – I was developing systems. I had a three-year contract, and I remember going back to Manchester to visit my parents and they said: 'Do you think you're going to stay in Switzerland?' And I said: 'I don't know, but I'm certainly going to stay until we see the first patient treated.'"

Thirty years later, Tony Lomax is still at PSI, now chief medical physicist at the Center for Proton Therapy.

The power of teamwork

Lomax remembers that in his early years, as the gantry project developed, there was a sense of urgency.

"What Eros and Hans wanted to do was to build this gantry-type system for delivering pencil beam scanning, where we could rotate the proton beam around the patient. We would create the first clinical facility in the world capable of using this technique to treat deep-seated tumours anywhere in the body. It was the first really practical and clinical implementation.

"We were very focused on treating the first patient, and there was even a little race with a parallel project being run at the GSI Helmholtzzentrum für Schwerionenforschung [Helmholtz Centre for Heavy Ion Research] in Darmstadt."

GSI had been established in 1969 as a German national research institute, with a central accelerator laboratory where researchers and university physics institutes from all over the world could conduct top-level research. In the early 1990s, GSI was also investigating a pencil beam scanning approach to treating tumours, but without a gantry and using heavy carbon ions, not protons.

"They knew what we were working on too, and there was a definite race," says Tony Lomax. *"It was a friendly rivalry between us to see who would be the first to clinically deliver pencil beam scanned treatments using charged particles."*

About ion therapy
Accelerated ions (atoms or groups of atoms with an electrical charge) are another type of high-energy particle that can be used for radiation therapy. Ions are heavier than protons, and very ballistically precise. Their medical applications have been explored in Austria, China, Germany, Italy and Japan since the 1990s. Using carbon ions results in very dense ionisation events in cells, and the damage caused by this cannot be repaired. This makes carbon ion therapy effective for radioresistant tumours. However, ion therapy requires very bulky and costly accelerators.

In many ways, the race was inevitable. The time was right for an explosion of innovation in radiation therapy because the technology required had become feasible. In the words of Martin Grossmann, a PSI physicist working in the IT Department who became involved in preparations for the gantry in the early 1990s: *"Implementing pencil beam scanning became possible when the technology was ready. Some of the ideas had been around for a while, but they needed the technology to catch up for them to be realised."*

But that wasn't the only ingredient for success. PSI provided an environment and ethos where medical doctors, physicists and engineers teamed up to pool their knowledge and experience. Gudrun Goitein was all too aware of the need to foster a collaborative working atmosphere.

"We didn't have our own offices, which is very unlike a clinic, where the bosses have their own secretary, their own room and so on. This intellectual fertilisation was of utmost importance, and still is.

"This was one of the keys to the success of the project. I remember the time when we hired Tony Lomax. He was so energetic and we knew that this young man was right for our project. So we spent endless hours together in front of the computer screen, explaining the radiotherapy side to him, the effects of radiation on different organs and body functions, why we used certain angles and doses. It was a very important characteristic of this project: for us all to understand why some things are possible and some are not."

It made for a highly productive, and occasionally intense, atmosphere.

"It was a small team, a compact team, and we were all very dedicated," says Lomax. *"We were discussing things all the time and it was exciting. Eros was the mastermind and the leader. In my opinion, he's a genius. But it's fair to say that I was young and possibly a bit arrogant when I came, and I was prepared to fight my ground when others in the department wouldn't. So there were certainly animated discussions sometimes. On the other hand, it was never a big issue."*

The gantry takes shape
Gantry construction started in December 1992, with the foundations being placed in a 75 cm-deep pit excavated in the floor of the NA Hall. The bearings for the rotating structure were mounted in the pit in the spring of 1993, and then the gantry supports – consisting of two large wheels connected by a welded box frame – were installed. Then came the rotating system with mounted beam-line elements, and the patient table structure.

Counterweights to the beam-line magnets were installed in the box frame as part of the supporting structure. Altogether, the beam line, the support with counterweights, the patient table and the X-ray devices added up to a total gantry weight of 120 tons. Initially, the whole gantry was attached to the NA Hall crane as a precaution, in case the weights had been misbalanced.

"The whole structure had been designed to be very rigid and very precise," says Eros Pedroni. *"The alignment of the elements of the beam line was planned to*

Figure 29. Eros Pedroni considers the pit into which the gantry will be installed. Pedroni remembers Gudrun Goitein asking him: "What happens if it doesn't fit?" "Then I will have to look for another job outside PSI," he said.

remain unchanged during gantry rotation of 185 degrees within tenths of millimetres."

But the gantry construction was a long and highly complex procedure which went far beyond installing mechanicals. As Eros Pedroni emphasises, in some ways, that was the easy part.

"A gantry is not just an iron block. People always think that if you build a gantry or something like it, the main issue is the mechanics and then you can do a beam line. In fact, it's the opposite. The most difficult part is the beam optics, because it's a multi-dimensional problem, involving position, direction and momentum. You have to bring the beam line over, say, 50 metres, just with the use of magnetic fields. That's something that needs a lot of engineers – the best professionals with expertise in fast changing of energy, the best engineers for the power supply.

"So this is the high-tech part of this business, and the competence of the Paul Scherrer Institute. The expertise we had to call on in this environment was ex-

Figure 30. The main support of the gantry is mounted.

tremely important – people like Ivo Jirousek, for example, who was responsible for remote control of the beam line. The people you have around is the success of the project – if you don't have them, then there is a problem."

Patient preparation

The large NA Hall at PSI had to house other projects apart from the gantry. Because of limited space, access to the proton treatment area was via a long straight corridor, with a sliding bunker door at the end isolating the gantry room during treatments. A control room, where technicians were to supervise dose delivery, was constructed on the first floor.

The pavilion for patient preparation outside the NA Hall started out as a small wooden barrack. Here medical personnel carefully positioned and stabilised the patient on the couch, which was in turn mounted onto a patient transporter.

Pedroni: *"A moving platform was installed at the front of the gantry. With the gantry in horizontal position and platform raised, it was possible to bring the patient-transporter system close to the gantry table and to couple the patient couch to the table of the gantry. After removing the transporter, the platform was moved down to allow the rotation of the gantry during treatment.*

"The position of the patient on the couch was checked with a CT scan in a CT room in front of the gantry entrance. The same CT machine was used before the start of therapy for taking the data used for treatment planning, including reference scout-view images (orthogonal projections). These were used for checking the position of the patient before each fraction. We would then transport the patient on the couch from the preparation room to the treatment area, where the couch would be connected to the gantry's patient table."

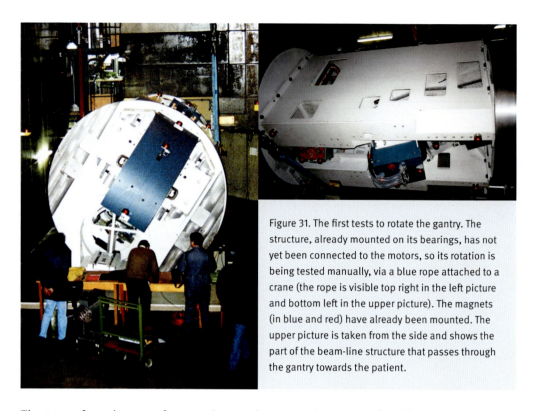

Figure 31. The first tests to rotate the gantry. The structure, already mounted on its bearings, has not yet been connected to the motors, so its rotation is being tested manually, via a blue rope attached to a crane (the rope is visible top right in the left picture and bottom left in the upper picture). The magnets (in blue and red) have already been mounted. The upper picture is taken from the side and shows the part of the beam-line structure that passes through the gantry towards the patient.

The team found ways of correcting patient position errors by offsetting the treatment couch. But there were still significant challenges. No matter how carefully a patient was positioned and stabilised, there was always the problem that living patients have active organs: the proton target might be a moving one.

Physicist Mark Phillips, on a sabbatical year at PSI from Berkeley, USA, investigated the problem of dose-homogeneity errors due to organ motion during scanning. His simulations, results and analysis were presented to PTCOG in 1991. Possible solutions included "repainting" the target tumour with additional scanning cycles and synchronising the scanning with the breathing phase.

"Organ motion was from the very beginning a major point of concern for starting our project," says Pedroni. As he pointed out in the paper published in 1995: *"Organ movements during treatment delivery are a challenge for any precision treatment. With the spot scanning technique, not only the shape of the edge of the field but also the dose homogeneity inside the target volume can be deteriorated by uncontrolled movements of the target during treatment."*

Addressing the organ motion problem would prompt many of the technical developments implemented in the PROSCAN project and Gantry 2 (see Chapter 6).

The animal treatment programme

The first test beam was sent through the gantry in April 1994. Work continued on testing the beam-delivery system and investigating any unexpected radiobiological effects that might arise due to the high dose rate at the proton spot. Detectors were used to understand and characterise the machine. Treating patients couldn't even be considered before there was certainty about safety and effectiveness. No one had ever constructed a machine like this before.

Gudrun Goitein: *"From previous experience with radiotherapy, we knew more or less what to expect when the human body receives a specific radiation dose. But when you suddenly have a so-called dynamic treatment that moves in the body, you need to make sure it has an effect in the exact area you calculated for. Even if you double safety systems and computer control of the beam, if the beam doesn't do what you predicted, and you're a bit off target, it's disaster. Once a beam is applied, the damage is there.*

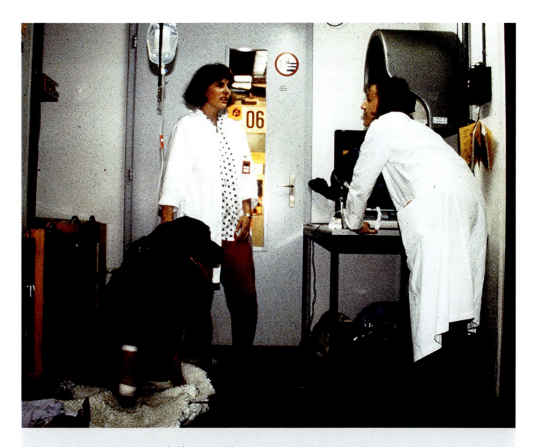

Figure 32. Barbara Kaser-Hotz (left) pictured during a pet consultation at PSI.

"So you don't want to be practising on human patients. That's where the animal programme came in. It was a very clever thing. It served us all well."

The proton therapy animal programme had actually begun a year before the gantry beam line went live, but it was instigated with the gantry project very much in mind. It was not "animal testing" as most people know it – trying out new treatments on healthy animals or animals in which illness or disease had been induced. Instead, this was a highly innovative programme using radiotherapy to try and cure pets of cancer.

Veterinarian Barbara Kaser-Hotz was central to the programme between 1993 and 1998, and was to become a major figure in the development of veterinary radiation oncology in Europe. At the time a senior assistant in the X-ray department of the Veterinary Faculty of the University of Zürich, she had trained in radiology and radio-oncology as a resident at the Veterinary Teaching Hospital at Colorado State University in the United States between 1987 and 1990.

In Colorado she had worked under the pioneering veterinary radiologist and radio-biologist Ed Gillette at a unit dedicated to providing radiotherapy for pets with tumours. Gillette, along with influential radiation oncologist Herman Suit, developed the "spontaneous animal tumour" model, recognising the value of examining naturally occurring cancers in animals to learn about human cancers.

"The basic idea was: why should we artificially induce tumours in animals, if there are so many dogs running around with spontaneous tumours that are in need of treatment?" says Barbara Kaser-Hotz. *"We can treat them with the most modern modalities and at the same time learn from those treatments for the benefit of everyone."*

Gillette's work had a huge effect on her. Back at the University of Zürich, she had worked with, and was inspired by, Professor Börje Larsson – previously a pioneer of proton therapy in Uppsala, Sweden, and a radiation biology researcher at PSI. She wanted to find ways to provide radiation therapy to animal patients, and he helped her.

"Professor Larsson was very enthusiastic and he brought me to PSI and introduced me to Hans Blattmann. Hans was instrumental in getting the whole animal programme up and running. He told me all about the pion programme, where there

had been too many side effects, and he was worried about the same thing happening again with protons. It was clear that going from mice to humans was too big a step. So Hans thought it would be ideal to use the spontaneous animal tumour model before the first patient was treated on the gantry.

"So I was introduced to everyone, and gave a lecture on what we planned. I have to say there was scepticism from some people. Luckily, Gudrun, a true animal lover, saw the potential and supported the idea. It was agreed that we should go ahead."

Gudrun Goitein: *"With the animals we could do two things. We could see whether the dose we calculated was tolerable and whether the effect was where we thought it was. But we could also test the whole procedure we were planning for human patients, ensuring we could get very good immobilisation in the area we wanted to treat, examining where the dose effect was. The animals were anaesthetised and immobilised in the same way as humans. They got a CT scan, treatment planning, a dosimetry – exactly the same steps we later did for our human patients. This way we could verify that the area of the body we needed to treat was in exactly the right position and receiving exactly the right dose. Learning these procedures was enormously important."*

With the gantry still under construction, the first animal treatments took place in the OPTIS facility in the early hours of the morning before eye patients arrived and after they had left in the evening. Although the ultimate aim was to test scanning on the gantry, the lower energy beam used for treating eye tumours provided opportunities to treat more superficial tumours in smaller animals, and practise immobilisation, anaesthesia and other procedures.

Kaser-Hotz: *"Treating pets with radiation therapy was totally new – not just to PSI but to the public in Switzerland generally. So one major goal of starting off in OPTIS was to get the people working at PSI used to the fact that they would see dogs and cats walking or being carried through the facilities. We were concerned about animal-rights people, and decided to have an open policy and explain what we were doing – that actually we were giving pets treatment they would otherwise never get.*

"People at PSI got used to us walking around with dogs. It took a while, but I think we became appreciated, and animals created a positive atmosphere. Over all the years, we never encountered any problems or had any negative press."

The first animal patients

The very first animal patient at PSI was a pet tortoise named Cleopatra, treated in March 1993.

Barbara Kaser-Hotz: *"The vet from the zoological department at Zürich University approached me because she didn't know what to do for this tortoise with a squamous cell carcinoma on the face. I knew that we could only treat very small tumours with the OPTIS beam, so Hans and I thought this would be a good start."*

Human doses had to be recalculated according to animal size.

"Human patients with ocular melanoma were treated with four fractions of 15 Gy, but the tumour of the tortoise was about half the size of its head, so I did not dare treat it the same way with such a high dose per fraction. So we treated twice daily with 4.8 Gy for four days. The tumour shrank and the tortoise survived well. Sadly she died of parasites a year later. But she really helped us get the animal programme off the ground."

Figure 33. Cleopatra the tortoise being prepared for proton therapy in the OPTIS facility.

Figure 34. Maugi the cat, who had a squamous cell carcinoma on her nose.

The second patient was a small cat named Maugi, with a squamous cell carcinoma on her nose. The owners had previously been told that the only option was surgical removal of the nose.

Kaser-Hotz: *"I approached the owners and asked them whether they would potentially be interested in trying an alternative – a new radiation therapy technique – on Maugi. After long discussions and trust-building, the family decided on the new therapy. I was very pleased, but at the same time extremely nervous. A cat loved by the family was being given into my hands. But I fully trusted Emmanuel Egger and his OPTIS team, and Maugi also received two fractions per day over one week.*

"All the animals were treated under light general anaesthesia to keep them from moving when the beam was on, which was obviously a major difference from treating humans. Although the beam therapy lasted only 90 seconds, perfect positioning took much more time. We intubated [inserted a tube down the throat to keep the airway open] *all the animals, including the tortoise, but used intravenous anaesthetic, as inhalation anaesthesia was not available and too complicated.*

"In those early days, we had to transport the animals from the veterinary hospital in Zürich to PSI, which is about a 45-minute drive by car, twice a day. At first, I loaded all the animals into my own car, but then someone very kindly donated a little Toyota wagon for the transportation back and forth."

In late 1993, Kaser-Hotz made the first of many visits to PTCOG meetings to report on the new animal proton therapy programme. In a paper jointly presented with Hans Blattmann, Gudrun Goitein and Emmanuel Egger in Cambridge, Massachusetts, she said that animals were being studied for acute and late tissue reactions and tumour control.

"Biologically, the selected spontaneous animal tumours behave very similarly to the same tumours in humans but often progress at a faster rate, and animals have a shorter lifespan, which means that results are available within a shorter period of time," she reported.

In all, 14 animals were treated using the OPTIS beam during 1993 and early 1994. Most were cats with squamous cell carcinomas, but dogs with melanoma of the toe and eye were also treated. Among the early discoveries was that the accelerated treatment schedules that were often necessitated by the limited hours available at OPTIS, delivering eight fractions morning and evening over four days, were actually more successful than conventional treatment over several weeks.

"They were like new after a few months, whereas those treated with the conventional scheme had many problems," remembers Emmanuel Egger.

Once the gantry beam line was operational, the animal programme moved on to larger pets. The first gantry patient was Jessie, a four-year-old Airedale terrier with a huge infiltrative lipoma (a tumour arising in fatty tissue) on her hind leg. Proton therapy offered an alternative to amputation.

Barbara Kaser-Hotz: *"I spent a long time explaining the procedure to Jessie's owners, and even showed them around the facility before they decided to enter Jessie into the trial. What finally convinced them to agree was the fact that if something went wrong within the treatment field, the leg could be amputated and Jessie would be okay. But I still felt a huge responsibility not only to Jessie but to her loving owners in Germany.*

"The start of treatment was scheduled for two o'clock in the afternoon, but things didn't run smoothly – as one might expect in a first trial. Things kept being delayed and I had to call the owners multiple times. Finally, Jessie was positioned in her wooden box and treatment could start. The gantry room was still full of cables and somewhat intimidating. We had EKG [an electrocardiogram] and a monitor around her chest to see whether she was breathing.

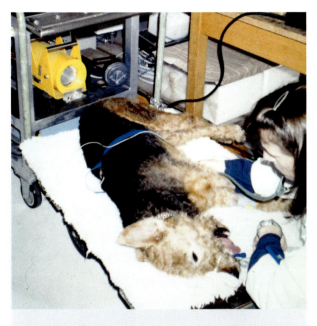

Figure 35. Jessie the Airedale terrier, the first animal patient treated on Gantry 1 in 1994. Here she is being prepared for treatment under anaesthesia.

"We were up in the control room looking down on Jessie intensely and constantly. We could not get to her to check that everything was okay, and we knew that if we stopped it would not just delay the treatment but other PSI researchers who needed to use the beam line. It was very stressful, and this first treatment lasted close to two hours because the tumour was huge.

"It was around three o'clock in the morning before I got back to the veterinary school with Jessie. It took quite a lot of convincing of the owners before they allowed me to do the second treatment. In total, we completed ten fractions of four gray over a month. By the end we had spent a lot of time at PSI and it was such a relief when Jessie's therapy was finally completed. It was all very stressful, but there was also a good feeling of having accomplished something which hopefully would bring radiation therapy forward."

Jessie survived for more than four years and, with her tumour stabilised, lived a good-quality life. The cause of her death was kidney failure unconnected with her tumour or therapy. The only side effects she ever experienced were permanent hair loss and hyperpigmentation.

Overcoming scepticism

Many more dogs followed. They had to be very carefully selected, with owners having to sign consent forms, agree to record their pet's condition following treatment and attend follow-up sessions for at least three and a half years. The team took on more and more complex tumours in difficult locations, such as the head and nose, where it was important to know that the treatment did not reach brain tissue. At the same time, they chose tumours that were unlikely to metastasise, because they wanted to be confident that the dogs would survive at least six months so that the follow-up could be for as long as possible.

Barbara Kaser-Hotz: *"Although, at the beginning, Swiss radiation oncologists trusted the physics aspects of proton therapy, they were worried about the response of tissue to this new treatment. That's why I reported frequently to the radiation oncology community here in Switzerland about the side effects, which we documented very carefully. I showed them many pictures of the dogs. The skin of animals is a perfect monitor for radiation side effects: you can see hair falling out then growing again, depigmentation then repigmentation. I believe these many reports, including numerous colourful animal pictures, helped to build confidence in the project."*

However, there was significant resistance to overcome from veterinary academia, university institutions and potential funders. At the time, veterinary science in Switzerland was more geared towards large and farm animals. Why should pets be treated with such a costly modality, they asked? Financial and institutional support for the animal programme was slow to arrive.

"My head at the university didn't really like me spending so much time working at PSI," says Kaser-Hotz. *"In the United States, Ed Gillette told me that the difficult part about introducing animal radiation therapy wasn't convincing the owners, but convincing veterinary colleagues, and he was right. Some people at the veterinary school thought I was crazy to irradiate pets. They said I should tell the owners to get a new dog."*

She encountered scepticism at PTCOG meetings too.

"The meetings were held twice a year, and it was very exciting to be the only veterinarian among all these smart people. I had to be careful not to give too much information about what we were doing, but it was also important to give informa-

Figure 36. Inside the Gantry 1 control room.

tion about the physics and radiobiology, particularly side effects, because there was considerable scepticism about this totally new scanning technique. Ed Gillette told me many years later that everyone 'in the scene' in the US thought that the Swiss were totally crazy. With these three movements, of the table and the gantry and the magnet-moved beam, they said: 'What happens if one of them stops?' I suppose everyone was happy that we were testing the technique on animal patients.

"In retrospect, it fills me with satisfaction that we never lost a dog during therapy, and never had to euthanase a dog because of side effects. One dog, Jimmy, lived more than ten years, and his tumour never came back."

The role Hans Blattmann played in convincing people of the value of the project should not be underestimated, says Kaser-Hotz. "He was the steady one, never giving up. He had the political skills and connections, and paved the way for the programme, often in the director's office talking about finances. Gudrun Goitein was also instrumental, defending the programme when scepticism arose."

The hard work of explaining, providing evidence and persuading did eventually bring results. Blattmann and the head of biological sciences at the University of Zürich convinced key university figures that the PSI studies would profit humans, not just pets, and the university began to contribute funds. Then came funding from the Swiss Cancer League and the Swiss National Science Foundation.

So the project could continue. Barbara Kaser-Hotz employed student doctors as assistants and bought anaesthesia equipment. New facilities were built to house the animals during their treatment days. "We got our own animal barrack at PSI, which was still standing until 2019. With time, we professionalised the treatments and brought in new equipment."

The equipment on the gantry also evolved: *"For the very first animal patients we had no imaging. There was a simple X-ray unit attached to the gantry head, so that's what we used to compare our positioning from fraction to fraction. We did fractionated radiation therapy from the beginning, and had a dose-escalation programme. About a year after we replaced the X-rays with CT scanning, which was not common at all in radiation oncology at the time. So we learned how to use the CT at PSI on the dogs, doing the computerised treatment planning and then matching the landmarks visible on the scan before every treatment.*

"The time I spent at PSI was one of the best periods of my life, because it was so inspiring to work with these clever doctors, biologists and physicists. Physicists are so analytical and work differently from people in the medical field. I loved to learn from them, and although I did not always fully understand, I completely trusted them. I was never pushed to treat an animal if I felt it was not ready or well prepared, and this was always respected by everyone. We had a lot of fun and many exciting discussions – a truly wonderful example of great team effort and collaboration."

The animal programme continued at PSI until 1998. For the proton therapy project, the knowledge gained was invaluable for pushing forward with human treatment. The scanning system worked.

However, the veterinary patients were treated without a completed safety-verification system. This work needed to be addressed. This was when physicist Martin Grossmann moved from PSI's IT Department to work on the software required for the scanning control and safety system.

By 1996, he says, the whole team had valuable experience in using the procedures they had devised. *"The animal programme gave us a lot of confidence in handling our machine, managing the treatment workflow, positioning the patient, taking X-rays and scans, bringing the patient into the treatment room, analysing the treatment protocol."*

The world looks on

Grossmann's new role was not without its pressures. The animal programme may have been a major stepping stone towards human treatment, but guaranteeing the safety of the system was now crucial before making that final significant step. Grossmann read extensively around radiation therapy, and it made him even more aware of his responsibilities.

"Sadly, in the early days of radiotherapy there were accidents. I read about one incident where several people were injured and one was killed because of a software bug. I was very aware that getting something wrong in the software could have very dramatic consequences."

The scanning technology was so new that there were no industry safety standards to follow. And the sceptical eyes of the world were now on PSI, curious whether such an adventurous approach would be safe.

Martin Grossmann: *"I didn't realise it at the time, but I learned afterwards that people were watching us, and basically letting us try it out and seeing whether it worked, ready to blame us if it didn't. Our management was very aware of this."*

There came another pressure in January 1996: a deadline. In late 1995, the PSI director, Meinrad Eberle, walked into Martin Jermann's office and told him that PSI's radiation oncologists had told him they would like to treat a human patient.

"He asked my opinion on this," says Martin Jermann. *"What about the risks? How would it affect the institute if it failed? We decided to start a review, bringing in an international group of experts including Michael Goitein, who would chair the review committee. They visited in mid-1996 to advise on what should be done before we started to treat the first patients and recommended some adaptations, upgrades and modifications – some to be done before we started the first patient treatment, others which could be implemented later."*

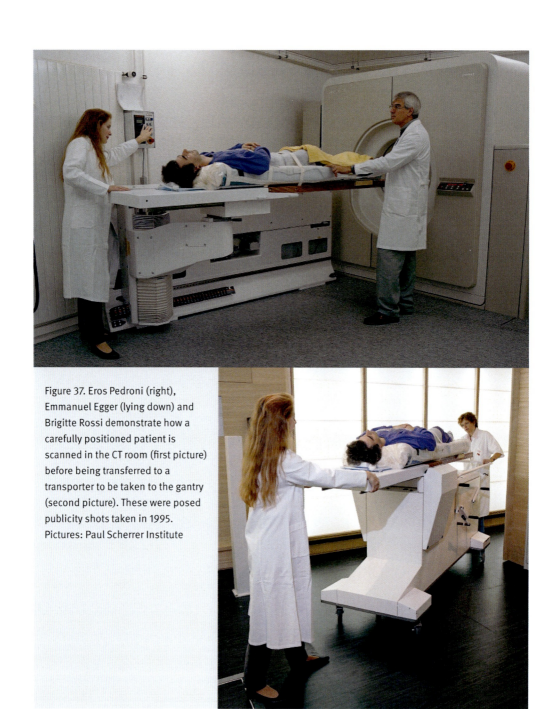

Figure 37. Eros Pedroni (right), Emmanuel Egger (lying down) and Brigitte Rossi demonstrate how a carefully positioned patient is scanned in the CT room (first picture) before being transferred to a transporter to be taken to the gantry (second picture). These were posed publicity shots taken in 1995.
Pictures: Paul Scherrer Institute

Eros Pedroni says he welcomed the review. *"It was hard and strenuous, but it was an opportunity to show that we did our best for guaranteeing maximal security. I presented the safety concepts and provided a written response to the report of the committee. The safety report was taken over and further worked on by Dölf Coray, who became responsible for the safety and quality assurance for the beam-delivery system of the gantry."*

If the recommended changes were made, human treatment could go ahead. In the meantime, proud of progress on the gantry project so far and buoyed by the success of the animal programme, the PSI directorate held an official inauguration of the gantry in January 1996. At the event, where invited guests toured the proton therapy facility, it was announced that PSI hoped to treat the first human patient before the end of the year.

"This was about the same time that I joined the team, and it meant there was significant time pressure," says Grossmann. *"The control system was basically working, but it was still lacking many features required for clinical operation. Implementing these features was only possible in close collaboration with the physicists on the project."*

Control and safety systems

Put simply, the role of the control system was to ensure that the treatment device strictly followed the treatment plan, explains Martin Grossmann.

"The basic concept for the control system was developed alongside the gantry hardware. Its main feature is redundancy: two independent computer systems know the therapy plan and apply it while continuously monitoring the application. One computer, called the Therapy Delivery System, takes the active role in setting the various machine parameters. The second computer, the Therapy Verification System, checks the correct setting of all parameters using independent physical methods.

"The Safety System is an additional system that supervises the treatment. Instead of verifying the parameters of a particular therapy plan, it constantly makes sanity checks to ensure the correct functioning of all components.

"We implemented 'logfiles', recording key parameters like dose and position for each scanned spot, and using particle physics tools for data analysis and visualisation. This turned out to be key for making the system work. At the beginning, we were suffering from frequent scan interruptions because the position-sensitive detectors of the gantry indicated a wrong spot position. The logfile analysis helped us understand that these interruptions came from spots with low dose where the measurement was not precise enough. Increasing the position tolerance for these spots, which was perfectly acceptable due to their low dose, made the system run smoothly through a complete scan."

Figure 38. Part of the Gantry 1 team pictured in 1996, including: PSI Director Meinrad Eberle (in jacket and tie); Martin Grossmann (in red); Emmanuel Egger (above Grossmann); Tony Lomax (left of Egger in green); Gudrun Goitein (below right of Egger); Hans Blattmann (extreme right). Eros Pedroni is not pictured because the photograph was taken as a gift for his 50th birthday. Picture: Paul Scherrer Institute

Grossmann and his colleagues were determined to make the system as safe as humanly possible. One accident could mean an immediate end to the whole project. Layers of "redundancy" were built into every level of hardware and software. If one system failed, there was another system to back it up, and if that failed, another one.

"But how many systems do you have to build in to switch the proton beam off if something goes wrong?" says Grossmann. "You need more than one, and the safety-standard norm is to say there should be two independent systems, like a bike has two brakes. We designed the system for the gantry so that there were five independent systems for switching off the beam.

"Maybe we were a little bit over-cautious. On the other hand, we have now treated more than a thousand patients on the gantry, and no one has ever suffered any damage as a result of a technical malfunction. So it has worked."

Rising to the challenge of minimising risk in patient treatment has been a source of enormous satisfaction to Grossmann.

"I do physics, develop electronics and write software because it's something I can do and like to do. But now I was doing it for a specific purpose: to get good results for patients receiving treatment. What better can you do?"

The first patient

In September 1996, the PTCOG meeting was held for the second time in Villigen. There were no formal presentations from PSI about the gantry pencil beam scanning project – although there were papers from GSI Darmstadt and the PSI OPTIS team. On the brink of a breakthrough but with the final major step not yet taken, this was not a time for reporting. And despite the success of all the preparations, the scanning project team's excitement was tempered with trepidation as the first human treatment neared.

Tony Lomax remembers that all of those contributing felt a responsibility. *"You can treat as many plastic phantoms and water baths as you like, but it is a totally different experience when you put a patient on the couch,"* he says.

"It was scary, to be honest. Some people in the particle therapy community were critical because they thought it was too dangerous. With pencil beam scanning, to make the delivery time reasonable, there has to be a high intensity at each of hundreds or thousands of spots – the number of particles per second has to be high. And people thought that was too risky: if the safety system fails, you can deliver 10 Gy within a second or two. At the same time, they were glad we were trying it out, not them. I think some people were a bit jealous because they could see that, if this worked, it could be the way forward."

It was one of Gudrun Goitein's guiding principles that procedures should be kept as simple as possible in early treatments. As experience and confidence grew, more complex tumours would be treated. *"The primary objective was to see that we were safe, and that our results met the standard of existing results from proton therapy using scattering, from Harvard in particular,"* she says.

So the first case provided palliative radiotherapy. The patient selected had brain metastases and required proton therapy as a supplement to palliative photon radiotherapy. He received the first human radiation treatment in the world using a

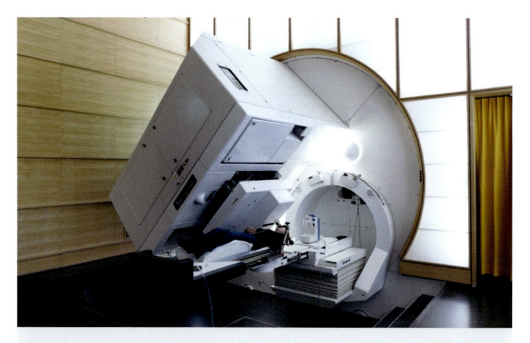

Figure 39. Publicity photo for Gantry 1. PSI Director Meinrad Eberle was keen that receiving proton therapy should be a good experience for patients, and commissioned Swiss architects Romero Schäfle to provide a pleasant décor in the treatment rooms for all the gantries. Picture: Scanderbeg Sauer Photography

pencil beam scanning system at PSI on 25 November 1996. The dose was delivered from one field (direction), and there were four treatment sessions in all. Everything went well.

Over the border in Germany, the GSI project didn't complete their first treatment using scanning technology until the following year.

Chapter 6

1997–2003
Project PROSCAN

Treating the first patient was a landmark for the Paul Scherrer Institute – top of the list of achievements in the PSI Annual Report for 1996. "We were able to treat the first human patient, successfully, in our unique cancer therapy facility, which destroys deep-seated tumours by means of precisely-guided protons," it announced.

But the first human treatment was not a source of celebration in itself. The sense of achievement was tempered by the fact that the patient had incurable cancer. The same applied for the very ill patients who immediately followed. During 1997,

Figure 40. The Gantry 1 team gather outside the proton therapy building.

nine further patients, referred from Swiss university hospitals, were treated – four with brain tumours and five with large tumours of the pelvis.

The early signs were hopeful: none of the primary tumours showed local progression after irradiation. Nevertheless, the success of proton therapy using pencil beam scanning would not be known until many patients had received it and their outcomes understood. Only time would tell whether scanning's high dose rates brought the benefits hoped for without severe side effects.

"There was certainly some sense of achievement with the early patients, but I guess it was a matter of wait and see," says Tony Lomax. *"Testing anything new in radiotherapy is worrying because it can take months or years to know if the treatment is a success or not. You can have acute toxicities such as skin reddening and hair loss during treatment, but the nasty toxicities come further down the road. So if a patient goes home at the end of the treatment, you can't just tick the box. The typical timeline to say whether a patient is cured or not is five years after treatment."*

Pedroni, Goitein and other team members reported on progress during the first full year of gantry operation in an article for Strahlentherapie und Onkologie.[14] Treatments in 1997 had, they said, demonstrated the practical feasibility of PSI's proton therapy facility. The beam period of 1997 was started very carefully, with only one patient at a time. Only at the end of the beam period were multiple fractions (up to six) per day applied, they said. The goal for 1998 was to treat up to about ten patients per day.

Gudrun Goitein's guiding principle was to proceed cautiously, with a view first of all to the "gold standard" in proton therapy set by Harvard using scattering. Its results for treating tumours at the base of the skull and large sarcomas of the trunk had to be matched.

"If you want to scientifically validate a method you must only change one parameter," says Gudrun Goitein. *"Our change was using dynamic spot scanning as opposed to passive scattering. We did not change the dose prescriptions per se*

---

[14] E. Pedroni, T. Böhringer, A. Coray, E. Egger, M. Grossmann, S. Lin, A. Lomax, G. Goitein, W. Roser, B. Schaffner, 'Initial experience of using an active beam delivery technique at PSI', *Strahlentherapie und Onkologie*, 175 (1999), 18–20, https://doi.org/10.1007/BF03038879.

or the dose per treatment session. We moved slowly – I would never have treated young children in the early days. We wanted to see that we were safe, and that our results met existing results– which happily they did – before moving on to the next step."

Technical improvements

The radiation medicine section was a creative and collaborative environment. Most components were not available from industry as they are today, so daily checks and constant innovation in hardware, dose calibration and dosimetry were required from PSI experts such as Dölf Coray, Terence Böhringer and Shixiong Lin. Safety and control systems for proton beam scanning continued to improve.

"We were working in a rather small team of around 15 people, with expert support from other departments at PSI, which allowed us to be very focused," says Martin Grossmann. *"Continuous feedback from operations and short implementation cycles made it possible to improve the system quickly, which was particularly important in the early years of the gantry.*

"The original platform choice for the control system was becoming a difficult legacy. In 1997 we ported the system to a new, sustainable platform – the same one adopted for the PSI's Swiss Light Source accelerator control system, exploiting the synergies that PSI offered as a research laboratory. The pencil beam scanning software was completely rewritten in a language accessible to more programmers at PSI."

Grossmann says he found the work on the gantry project immensely satisfying. The main work was not done through meetings, but wandering into offices and having conversations. *"There was always a specialist at PSI to call on if you had a problem, whether it be with electronics, software, magnets or particle physics.*

"When you're developing control systems, you're working with a lot of software and electronics, but they have a connection with the real world: they open a valve, or a beam gets switched on, and you see the counter counting upwards. I find this fascinating."

It wasn't only control systems that were developing. Work on a revolutionary new approach to treatment planning for proton therapy was progressing apace.

A major innovation in treatment planning

Since his arrival at PSI in 1992, Tony Lomax had been adapting and improving the treatment planning systems first envisaged by Pedroni and Stefan Scheib, and named by them "multiple field optimisation".[15] This utilised the rotation of the gantry to look at treatment planning in a new, three-dimensional way.

"I had overlapped with Stefan Scheib for about a year and a half," says Tony Lomax, *"and when he left I took over his work, doing the coding to optimise what he had done, but basically the physics had been put in by Stefan, supervised by Eros."*

Treatment of most deep-seated tumours requires radiation to be delivered from more than one angle (field) to better focus the radiation dose at every part of the cancer. The difficulty is ensuring that every corner of the tumour receives the same, optimal, dose. It was clear that multiple field optimisation held immense potential here. For Eros Pedroni, it had been one of the motivations for building the gantry – it would allow the delivery of several inhomogeneous dose distributions and their combination into a total optimised dose.

In the 1990s, similar approaches were being proposed for planning treatment with conventional radiotherapy using photons, most notably by Thomas Bortfeld in Heidelberg. In conventional radiotherapy, the technique was called "intensity modulated radiation therapy" (IMRT). IMRT uses advanced computer programmes to calculate the direction of the field and the beam intensity from each field, for each treatment, so that the full volume of the tumour is covered with the optimal dose.

Lomax had watched IMRT for photon therapy develop as the power of computers increased. Unlike scattering approaches, three-dimensional scanning provided the ideal technology for an intensity modulated approach to treatment planning, ensuring that the entire volume of a tumour was "painted" with a uniform dose. The efficacy of three-dimensional IMRT using protons could far exceed that of conventional IMRT using photons.

---

[15] S. Scheib and E. Pedroni, 'Dose calculation and optimization for 3D conformal voxel scanning', *Radiation and Environmental Biophysics*, 31 (1992), 251–56. https://doi.org/10.1007/BF01214833.

"With these approaches you've basically got to do two things," says Tony Lomax. "First, you've got to define the beam angles for bringing your radiation to the tumour. Typically you would have seven to nine fields equally spaced around the patient. Second, you need an optimisation algorithm to convey a homogeneous dose across the tumour.

"But things aren't often that simple in radiotherapy. A brain tumour might be wrapped around the brainstem. Or maybe a head and neck tumour is partially wrapped around the spinal cord. So it's an interesting balance: getting as much dose to the tumour as you can, while reducing the dose to neighbouring structures such as the spinal cord. The power of IMRT is that once you've defined your angles, you can set your target for the optimisation algorithm: you can specify that you want this much dose delivered to the tumour and you don't want any dose to the rectum or spinal cord.

"So the algorithm will calculate the beam dose from each direction required to achieve that goal, literally sculpting the high dose to the tumour and scooping out the dose for areas of normal tissue. People call this 'dose sculpting'."

What Lomax developed was a system for optimising not just one field at a time, but all the fields simultaneously. In this system, every single proton spot from every single direction would be optimised as part of the big picture of how best to carry a uniform dose to the whole tumour. As Tony Lomax wrote later: "By IMPT, we refer to the delivery of a set of individually *in-homogeneous* proton fields, which when combined, deliver a homogeneous dose across the target volume."[16]

The prospect of dose sculpting using the power and accuracy of proton scanning technology was an exciting one. It could be far more finely targeted than anything using photons.

"The way I looked at it, this kind of optimisation was fundamental. I didn't do a lot, to be honest. All I did was just put the code together – probably the first in the world – which allowed us to define all the pencil beams in three, four or five fields, and optimise them simultaneously.

---

[16] A. J. Lomax, T. Boehringer, A. Coray, E. Egger, G. Goitein, M. Grossmann, P. Juelke, S. Lin, E. Pedroni, B. Rohrer, W. Roser, B. Rossi, B. Siegenthaler, O. Stadelmann, H. Stauble, C. Vetter, L. Wisser, 'Intensity modulated proton therapy: a clinical example', *Medical Physics*, 28:3 (2001), 317–24. doi: 10.1118/1.1350587. PMID: 11318312.

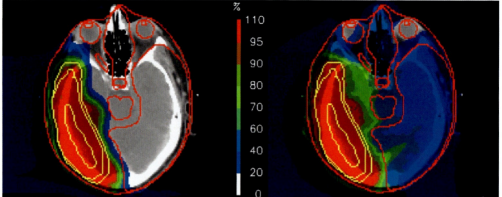

Figure 41. CT scans of a head from above, comparing the dose distribution of IMPT (on left) and IMRT when treating a meningioma (outlined in yellow on left of skull). The red outlines indicate organs at risk. The colour scale indicates percentage of dose delivered to the tumour. The lack of coloured areas beyond the tumour on the left indicate the precision of IMPT compared with IMRT.

"For example, if you were treating brain cancer you might have a field which passes through the brainstem. If you're treating with a single field you simply can't reduce the dose to this because you have to treat the tumour behind it. But if you are coming from two or three other directions, you can switch off all the pencil beams coming through the brainstem, and adjust the dose accordingly from other directions. Multiple fields are optimised and each field knows what the other is doing."

After extensive testing and comparison of different techniques of intensity modulation, Lomax introduced his work to the world in a paper written in 1998, and published a year later in the journal *Physics in Medicine & Biology*.[17] The technique was to become known as IMPT – intensity modulated proton therapy – and it has since become the standard treatment planning approach used in proton centres around the world.

The first results

PSI's Division of Radiation Medicine had become part of a Life Sciences Department described in the 1998 Scientific Report as "a small 'microcosmos' within the PSI, aiming to perform high quality research in biosciences focused primarily on

---

[17] A. Lomax, 'Intensity modulation methods for proton radiotherapy', *Physics in Medicine & Biology*, 44 (1999) 185–205. https://iopscience.iop.org/article/10.1088/0031-9155/44/1/014.

oncology." The whole department took up around 13 per cent of PSI's total budget, but radiation medicine a fraction of that.

The first results from pencil beam scanning on the gantry were beginning to come through.

In the 1998 Scientific Report, Goitein, Lomax and Blattmann reported that a further ten patients had been treated on the gantry, bringing the total to 20. The patients had malignant lesions in the brain, skull base, orbita, soft tissue of the shoulder, retroperitoneum and pelvis. Seven had been palliative cases, four of whom had died of their metastases. The remaining 13 appeared to be locally cured: "So far we have not seen a local tumor progression, even not in those four patients who died from distant metastases." Overall, the evidence was that radiation therapy using spot scanning protons was "effective and well tolerated".

The radiation medicine team also reported to the PTCOG meeting in April of that year that pencil beam scanning could bring benefits to patients with irregularly shaped tumours beyond a highly conformed dose. All nine patients in the study tolerated proton therapy very well, even at high doses.

The end of the animal programme
The animal programme, meanwhile, was continuing its dose escalation programme to increase understanding about the way tissue tolerates doses administered over different time periods. Studying the effect of proton therapy on nasal tumours in dogs was of particular interest. In dogs, the olfactory bulb extends very close to the eyes and brain, making radiation treatment without severe side effects extremely difficult.

"With traditional radiotherapy, the eyes and brain are always in the field," says Barbara Kaser-Hotz. *"Our dogs with nasal tumours treated with protons did not live significantly longer than traditionally treated dogs, but they had far fewer side effects. We never had any out-of-field recurrences, which was the big fear at the time."*

Animal treatments on the gantry were now scheduled either early in the morning, well before human patients arrived, or late at night – sometimes at three in the morning. However, the number of humans being treated on the gantry was increasing, and it was more and more difficult to justify treating humans on a machine that was also used for dogs. Despite the valuable information the programme provided, animal treatments only continued until 1999.

For the 30 or so dogs treated with proton therapy and monitored during the programme, the outcomes were good. Median survival was 385 days and many died of causes completely unrelated to their cancer or treatment.

For Barbara Kaser-Hotz, the project was the springboard to introducing the first radiation-oncology service for pets in Switzerland, after the Veterinary Medical School at Zürich University received a donation to buy an accelerator of its own. Today it is common across Europe for dogs to be treated with radiation to cure or stabilise cancer.

The ending of the animal programme coincided with both a new awareness of the proton therapy project from the outside world, and a growing appreciation at PSI that its radiation oncology work held an importance that extended beyond speculative research. Early results indicated that PSI's innovative approach had genuine potential to halt or cure cancers that had previously been untreatable. This wasn't just about scientific discovery but about curing human beings.

"The gantry was attracting a lot of attention from the public, politicians and international experts," says Eros Pedroni. *"It was increasingly playing an important role as a 'visiting card' for the institute, and the director of PSI, Professor Meinrad Eberle, was genuinely interested in its progress."*

An unsatisfactory situation
While the potential of pencil beam scanning was now clear, those involved with the project were also increasingly aware of inadequacies in the current facilities.

For example, some gantry positions made access to the patient difficult. Eros Pedroni knew that a false floor below the patient table would have solved the problem. *"Unfortunately, we did not have enough money, resources and time to provide a solution at the beginning of the gantry project, and after the system was in operation it was too late to provide a big change in mechanical structure,"* he says.

More frustrating were problems associated with the beam line. Every winter, the 590 MeV large ring cyclotron that supplied protons for projects throughout PSI, including radiation therapy, had to be closed down for four months for maintenance and upgrades. That meant no protons for treatment, so all patient referrals during that time – and during a "ramping down" and "booting up" period – had to stop. In all, the gantry was fully operational for six to seven months of the year.

Not being able to take patients during shutdown periods severely hampered recruitment from Swiss universities and limited the number of people who might benefit from proton therapy. University hospitals needed to be able to refer patients when they needed treatment, not when the accelerator was available. The restricted treatment time also made it difficult to keep to the scientific protocols required for clinical trials.

Even when the beam line was available, it was not always of a sufficient quality to enable proton therapy. The gantry facility was fed with a proton beam that had been split from the large cyclotron's main beam. But the operation of the splitter had been deteriorating since the proton current in the cyclotron had been increased to supply another PSI project – the spallation neutron source, SINQ. This resulted in intensity spikes in the beam for the gantry, meaning it was sometimes too unstable to deliver treatment. Safety systems were regularly kicking in to shut down the beam.

Eros Pedroni: *"The operation of the medical facility was in these situations frustrating: waiting for hours for the beam, interrupted treatments, terminating treatments late in the evening, shifting patient treatments to other days. It was becoming clear that the proton therapy project, the cyclotron operation with a very high current and the ongoing modification for SINQ were competing developments. The 'parasitic' use of the beam for therapy by stealing a little of the main beam was not a satisfactory solution for working with patients."*

The dream was a dedicated accelerator for proton therapy. That dream was about to take shape.

A decision of "epochal" importance
By spring 1998, it was clear that a choice had to be made about the future of radiation medicine at PSI. Having pioneered the proton scanning technology, and with over 20 patients treated so far, the institute had to decide whether the project should end or whether to develop it further so that patients could be treated more effectively. Gudrun Goitein had made clear to Martin Jermann how difficult it was to deliver a reliable treatment programme on the gantry, and that this now had

implications for getting patient referrals, conducting high-quality research and furthering the reputation of PSI.

"There was pressure on us from the clinicians, and I totally agreed with them," says Jermann. *"They pushed us to move as fast as possible to have a dedicated accelerator and then, in a later step, a new gantry."*

PSI's director, Meinrad Eberle, decided to set up a group to review proton therapy at PSI, headed by Michael Goitein, who had recently been appointed scientific adviser to the Division of Radiation Medicine. The review group included radiation oncology and medical physics experts from the university hospitals of Zürich, Geneva and Lausanne, representatives from the OPTIS and gantry programmes and Martin Jermann. Its mission was to "evaluate all reasonable options open to PSI with regards to the continued medical use of proton beams".

In October 1998, the review group presented its findings to a group of Swiss dignitaries and radiation oncology experts at a meeting in Bern. Michael Goitein presented three options for moving forward.

The first option was to complete research and development at the gantry but end large investments, letting OPTIS run down (the Injector 1 cyclotron that provided its protons was nearing the end of its lifespan) and transferring knowhow to the University of Lausanne.[18]

The second option was to continue both the gantry and OPTIS programmes, but with a new dedicated medical accelerator. In the light of an additional proposal from Eros Pedroni, there were two ways to pursue this option. PSI could continue treatments and research on the existing gantry. Or, given the increasing numbers of patients that could be treated with a new dedicated accelerator, it could start to develop a technologically advanced second-generation gantry, well suited to a hospital environment, which could be developed into a commercial product.

The final option was to end all investment in proton therapy at PSI, transferring all the technology, learning and expertise to a new hospital-based facility somewhere in Switzerland.

---

[18] See Chapters 1 and 3 for more information about the injector cyclotron, which accelerated protons to the main 590 MeV cyclotron as well as providing the protons for the OPTIS facility.

"We discussed these three options and concluded that it would be a huge risk to transfer all this technology and build an advanced gantry prototype in a hospital environment from scratch," says Martin Jermann. *"So we decided on option two, building a new accelerator and building a second gantry at PSI. This was supported by the group of Swiss decision-makers in Bern, and I got the mandate to look into how this could be implemented. How could we extend the programme? Where could we build the dedicated accelerator and the second gantry?"*

Jermann had to report back to the directorate by February 1999, presenting a strategy, business plan and development proposals. This was the beginning of what became known as PSI's PROSCAN project. The project was to mark the transformation of PSI's Radiation Medicine Division into a new Center for Proton Therapy – a facility that would truly serve patients, not just the advancement of science and technology.

It represented a massive vote of confidence in the proton therapy technologies developed at the institute, not least from PSI Director Eberle.

Eros Pedroni: *"Eberle's interest in proton therapy brought a decision of epochal importance for PSI and the future of proton therapy in Switzerland – the decision to acquire a dedicated accelerator for proton therapy."*

A commitment with conditions

Jermann returned to the directorate to present his proposals. A site for the new medical cyclotron had been identified inside the NA experimental hall, alongside the existing gantry. The plan was that, in stage one, the experimental hall would be prepared, the cyclotron built and a beam line connected to the existing gantry, so that year-round operation could begin as soon as possible. In stage two, the OPTIS facility would be moved into the experimental hall next to the Gantry 2 site. In stage three, when Gantry 2 was complete, it would be connected to the new medical beam line.

His business plan estimated that the cost of the PROSCAN project would be between 25 and 30 million Swiss francs.

Martin Jermann: *"Gudrun Goitein and I were at the February meeting, in front of all the heads from the different departments of PSI, who were part of the directorate which had to make the decision. There were ten of them and I had to convince them that it was worth making this decision. I knew that each one tended to look after their own interests, and spending in other departments could mean they got less for their own.*

"When we started proton therapy in 1988, the budget for the radiation therapy group was around 1 per cent of the total budget of PSI. It was clear that the investment required in my proposals would bring that proportion up to around 5 per cent for coming years."

The directorate agreed to his proposals, but only by one vote, with support from Eberle and one department head. The rest of the heads remained neutral.

"I remember that after the meeting, Gudrun Goitein said to me: 'You have no active support from the directorate!' I said: 'It's not so bad!' I knew from experience how it worked. I told her that though it seemed negative, if the heads of department didn't say anything, that was a lot of support. In the years after, there was never criticism from the other department heads."

But the support was not unconditional. The directorate decided that one third of the cost would come from PSI's operational budget, one third from sponsorship and donations, and one third from licensing the technology to industry. *"I had no idea if we could reach that, but I agreed and started working on it,"* says Jermann.

Another condition was that the treatment cost had to be covered by health insurance. Given the novel nature of this type of proton treatment, this proved easier said than done. It would involve the Swiss government making a decision.

"It was definitely a risk for PSI to be investing so much, because we did not know how the world would respond to spot scanning," says Martin Jermann. *"It was my job, together with Eberle, to go to foundations and sponsors and find two thirds of the money. Eberle and I were both convinced we could do it, because this was to do with treating cancer. But it was nevertheless a hard job."*

New landmarks in treatment
Meanwhile, the medical physicists and clinicians working on the gantry continued to break new ground. The year 1999 saw the first patient in the world being treated with the new technique of intensity modulated proton therapy (IMPT).

"I remember it very clearly," says Tony Lomax. *"It was a 34-year-old man with a lumbar spine chondrosarcoma, slightly wrapping around the spinal cord, which created a classic treatment planning problem. If we had used the old technique we could have treated it very well, but we wanted to avoid beams through the heart and lung because of possible movements in these organs causing dose errors."*

Treatment "overshoot" into the spinal cord also needed to be avoided. It was a perfect case for the detailed "dose sculpting" of IMPT.

Tony Lomax: *"We did a plan dividing the target into different sections and then, using this multiple-field optimisation, we could get a plan which looked very similar to a single-field treatment but didn't have any Bragg peaks of high dose stopping against the spinal cord. We could show that this IMPT plan was much more sensitive to potential range errors and made it much safer."*

As at 2022, the patient is still alive.

The year also saw the first children being treated with proton therapy. The positive experiences of positioning and treating adults encouraged Gudrun Goitein to include three children in the 1999 treatment programme. All were aged between 7 and 11. Two had inoperable brain tumours and the other a spinal tumour. No general anaesthesia was used, and all the children kept their position well and completed their treatment course.

The radiation medicine team reported in a PSI internal bulletin: "There were no acute side effects except expected local hair loss and grade one skin reaction . . . Daily positioning was very satisfying and well tolerated. Proton therapy using spot scanning beams has shown to be a non-traumatic treatment modality for CNS lesions in paediatric patients."

By the end of 1999, over 40 patients had been treated using spot scanning at PSI. Within the facility there was a sense that something significant was happening, on both a human and a scientific level. Damien Charles Weber, who had arrived at PSI as a young radiation oncologist in 1998, remembers he felt like a child at Christmas during his early years.

"It was a wonderful playground where everything was built from scratch," he says. *"It felt as if I, along with everyone else working there, could make a significant con-*

tribution to the management of cancer patients. I was determined to demonstrate that proton therapy could be of value to patients."

And as contact increased with real patients whose lives hung in the balance, it was difficult for team members not to feel personally involved. Tony Lomax remembers many former patients from these years.

"You certainly get a real sense of elation when things go well. I still try to get to patient-review meetings when we discuss past patients, and I get a real sense of achievement from every patient who has been cured at PSI.

"I remember a few years back we were discussing a French girl who we treated in the early 2000s. She was around 11 at the time. While she was on the gantry, our radiation technician used to read a French book to her over the microphone from the control room, to keep her calm. When it came to her final fraction, I said I'd take over the reading, as a joke. I didn't speak French, and because of my broad Mancunian accent she spent most of the time laughing – which wasn't the best thing for accurate treatment, to be honest, so I stopped. But recently I learned that she's got a family now, and has gone back to horse riding. Anything like that gives you a sense of achievement. The other side of it is that if you get a patient coming back with treatment toxicity, you ask yourself why."

Standing alone
At the turn of the century, the proton therapy programme at PSI stood completely alone. When Eros Pedroni spoke at the 7th European Particle Accelerator Conference (EPAC) in Vienna, Austria, he noted (in the written version of his paper): "Within the field of proton therapy the competition is between the established traditional beam delivery techniques using passive scattering and the new methods based on magnetic beam scanning. The innovative developments are compact gantries dedicated to beam scanning capable of delivering 'intensity modulated proton therapy'."

There was only one such compact gantry in the world: the one at PSI. Gantries using passive scattering had now extended beyond the hospital-based facility at Loma Linda to two facilities in Japan, in Kashiwa and Tsukuba. But only PSI's Gantry 1 used active scanning and the new technique of IMPT.

"Despite many uncertainties, proton therapy is steadily gaining in socio-economical importance," wrote Pedroni. "About ten hospital-based proton or heavy ion therapy facilities are now under construction or already available in the USA and Japan. Europe is waiting."

Would PSI's non-hospital-based facility show the way? History tells us that it did, but it was more than ten years from PSI's breakthrough first gantry treatment before another facility started treating with pencil beam scanning. The scientific world remained to be convinced that scanning had advantages over scattering.

"As far as PSI was concerned, the battle was won and scanning was the way to go," says Martin Jermann. *"But outside, people were still critical. Bernard Gottschalk, for example, who helped develop passive scattering at Harvard, thought that scanning was a nice idea, but it was too complex and expensive to be taken up widely. At a time when even conventional radiation was believed to be expensive, some people thought it was going in the wrong direction. Eros and I, however, thought scanning would not be more expensive than scattering in the end."*

None of this made Martin Jermann's task of raising funds and covering costs any easier. When it came to getting health insurance to cover the cost of proton therapy, the immediate barrier was that treatment costs would be two and a half to three times as expensive as IMRT with photons.

In the light of PSI's challenging condition that treatment costs had to be refunded by health insurance, it fell to Martin Jermann and Gudrun Goitein to apply to the Swiss Federal Office of Public Health for coverage.

"The first reaction from the Office of Public Health was: 'We have to look at this,'" says Martin Jermann. *"We had less than five years' experience with proton therapy, and the only evidence was from scattering technology in the United States, with quite limited indications. So the Office of Public Health said first you must bring the evidence about results and toxicity, then we can make a cost comparison. So we went away and wrote the application. We also needed to show that the higher cost was justified. We were very grateful to Professor Lütolf, director of radiation oncology at the University Hospital in Zürich, who gave us strong support."*

Potential donors, too, took some convincing. *"They were worried that this was a long-term project, requiring investment for an operation period of maybe 30 years.*

I was told that there were a lot of new cancer therapy drugs in development. How could we be sure that radiation therapy would still be needed 20 years from now? I was convinced, but it was a different matter convincing them."

The new cyclotron

Meanwhile the groundwork for the PROSCAN project began. Eberle and Jermann organised several international reviews to examine what industry could contribute to PSI's plans for the gantry, beam lines and accelerator. Options for the layout of the extended proton facility were evaluated, budgets elaborated, negotiations with potential industrial partners begun and specifications drawn up.

Figure 42. An artist's rendition of the superconducting cyclotron, to be designed and built by ACCEL in close collaboration with PSI. Picture: IOP Publishing (reproduced with permission, all rights reserved).[19]

The new accelerator was to be the heart of the new proton therapy facility, and it had to be perfectly suited to the job. It needed to take up relatively little space and deliver a stable beam which could be modulated to different intensities within milliseconds. An accelerator with a fixed energy would be acceptable, since fast energy changes could be provided with a degrader.

It was decided that, despite PSI's long experience designing and building cyclotrons, this option was not feasible given the demands on human resources for other big PSI projects. On the other hand, buying an accelerator "off the shelf" was not ideal either: there were fears that it would be difficult to adapt to PSI's own requirements.

---

[18] Published in: J. M. Schippers *et al.*, *Journal of Physics: Conference Series*, 41 (2006), 61–71.

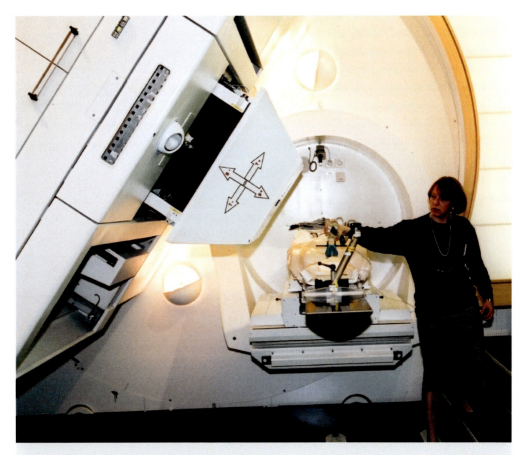

Figure 43. Gudrun Goitein demonstrates Gantry 1 during the press event in 2000.
Picture: Paul Scherrer Institute

A way forward was to find an external company to build a prototype to PSI designs and specifications, with PSI and the industrial partner then able to commercialise the product under a licence agreement. This was a good solution, since PSI had a clear idea of the type of design it would like. This was based on its extensive past experience with cyclotrons, and also on its knowledge of an innovative design of compact superconducting cyclotron created by Professor Henry Blosser, director of the National Superconducting Cyclotron Laboratory at the Michigan State University in the USA. Blosser's design would produce a stable beam, be relatively cost-efficient and take up little space. It would form the basis for the PSI cyclotron design.

Negotiations began with potential industrial partners who could manufacture the accelerator. Offers came in from ACCEL GmbH, Hitachi, IBA and other companies. ACCEL provided a detailed technical proposal.

On 14 November 2000, PROSCAN became public knowledge when PSI held a press conference to formally announce the project. Publicity in the Swiss press followed. "Reporters from the media, together with representatives of sponsors and the medical profession, were deeply impressed with the pioneering work of PSI, and this was reflected in the favourable response in the media," said the PSI Annual Report for 2000. "PSI now intends to develop it for application in hospitals so that the largest possible number of cancer patients can benefit from it . . . Prominent sponsors have already offered their support, and PSI hopes to find others."

By the end of 2000, PSI had treated 72 patients on the gantry, 60 of them with the intention of cure. Of these, 24 had chordomas and chondrosarcomas of the skull base, vertebral column and sacrum; 21 had tumours of the brain, soft tumours, sinuses, head and neck and prostate; and 15 had meningiomas, tumours of the membranes surrounding the brain which can lead to serious symptoms such as vision loss.

"In one case, now 18 months after irradiation, a most impressive response to therapy can be reported," said the annual report. "A female patient with almost total loss of vision due to a large meningioma between the eyes now has almost normal visual acuity."

A total of 236 eye patients were treated in the OPTIS facility in 2000. "The excellent results of proton therapy for melanomas of the choroid located at the optic fundus are yet another highlight in cancer therapy: 98 per cent tumour control is a result equal to the best that can nowadays be achieved in the fight against cancer."

As new leader of the proton therapy programme, Martin Jermann decided he needed to know more about the clinical side. *"I was previously focused on the physics and technology. Now I wanted to know every step of treating the patient, from first to last contact – to find out what you did as a radiation technician, a physician, a medical physicist and so on. Gudrun Goitein was very open to this, and she asked one lady patient if she minded if I was present during her treatment.*

"So this is how I spent my summer vacation, learning every step and every speciality – making a therapy plan, making a mould, examining the CT scan – finding out all the things I needed to know about how the patient feels and what the clinician and the medical physicist does."

Agreements made, contracts signed

PROSCAN and the planned expansion of PSI facility got under way in May 2001, when the Swiss government agreed to reimburse treatment for five years. Given the current lack of evidence about the cost-effectiveness, there was a stipulation that PSI should report annually about outcomes. The continuation of reimbursement at five years would be conditional on good results.

Martin Jermann: *"I was happy with that, because I was convinced we could demonstrate success – especially since the experience from Harvard was that there was less toxicity and better outcomes, particularly for cancer in the spinal cord. The reason was that you can give more dose to the tumour without disturbing the rest of the spinal cord, and we knew we could achieve this better with scanning than with scattering."*

Now PROSCAN began to gather pace. The expansion of the proton therapy facility officially began in June 2001 when a contract was signed with ACCEL to construct

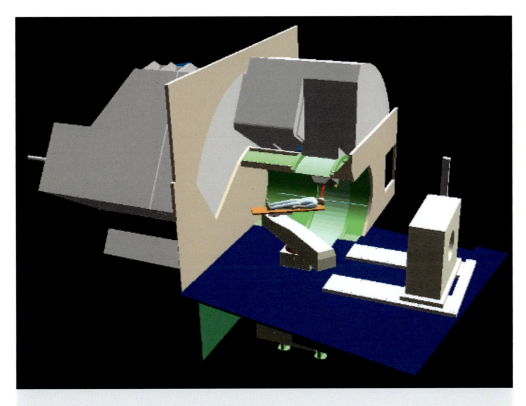

Figure 44. An early visualisation of Gantry 2 by Eros Pedroni showing the patient table, the false floor in front of the gantry (in blue) and the sliding CT scanner (see Chapter 7).

and deliver the dedicated cyclotron according to agreed specifications. The cyclotron was to be called COMET, standing for Compact Medical Therapy cyclotron.

"With this decision and the insurance companies signing up for five years, we could sign the contract for the accelerator and start the collaboration with industry to build the whole system," says Martin Jermann.

As overall manager of the project, Jermann worked to a steering committee headed by the PSI director. He organised international reviews to inform gantry specifications and the development of gantry control systems. Eros Pedroni led on the general technical specifications of the facility. Gudrun Goitein provided clinical input, as operational and medical leader of the proton therapy centre. And engineer Jürgen Duppich, head of the Department of Technology, Coordination and Operation, took charge of assembling the components and systems of the PROSCAN facility.

"Jürgen was the person responsible for managing the whole infrastructure, dealing with companies, installing the cyclotron and so on," says Martin Jermann. *"He was extremely important."*

When Jermann and Pedroni had early discussions about the project, they estimated it would take around five years to construct Gantry 2. This was the timeline that Martin Jermann outlined to colleagues in 2002. In fact, from the agreements being signed to the first patient being treated on Gantry 2, it took ten years. *"I was too optimistic,"* says Jermann. *"There was too much technological development still to do. It was an advanced new design, and we had other projects running in parallel at PSI, such as building complex instruments at SINQ and the Swiss Light Source, which were competing for resources."*

Some delays were caused by problems with the cyclotron construction. ACCEL had planned this to be one of the first cyclotrons of its type to be completely designed and developed using computer technology. Unfortunately, simulations showed that the iron used to construct the cyclotron was too impure to fit the computer modelling. Sourcing the kind of iron that had been used in the design simulation proved difficult, and the cyclotron was in the end delivered a year later than expected.

However, the collaboration between PSI and ACCEL worked well, and there was growing excitement about the innovative cyclotron. The work of the Accelerator Division and the commissioning work of Marco Schippers were crucial to this. As the PROSCAN team reported to the steering committee in 2003: "This 250 MeV superconducting cyclotron is under construction at the company ACCEL Instruments GmbH and shall be operational at PSI in the second half of 2004. PSI contributes to the development of the new accelerator, for example by analysing the static and dynamic 3D magnetic and electric fields with the in-house developed particle tracking program ..."

"Great attention was paid to the goal to provide high speed in changing the energy of the beam during scanning (to provide multiple target rescanning – for the treatment of moving targets with scanned beam – in order to cope with the organ motion problem)." Key to the accelerator design was an innovative vertical deflector plate installed in the cyclotron. It would allow the intensity of the extracted proton beam to be modulated in a tenth of a millisecond.

Swiss funders were coming round to the idea of supporting this adventurous project that could transform cancer treatment. In 2002, the Parliament of the Canton of Aargau agreed to loan PSI five million Swiss francs to support PROSCAN.

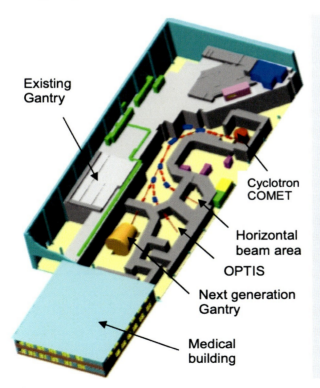

Figure 45. A "tentative layout" of the new proton therapy facility as envisaged in a 2003 PROSCAN status report. It read: "The existing Spot-Scanning Gantry will be connected through a new beam transport system to COMET. A new treatment area is foreseen for a new Gantry with additional features and improvements compared to the present one. Two additional horizontal beam lines, one for eye treatments (transfer from the Injector 1 of the OPTIS program) and one area dedicated to experiments are included." Hans Reist of the accelerator group was responsible for designing the layout of the beam lines to the different treatment areas.

Providing the evidence

At the May 2002 PTCOG meeting in Catania, Italy, Gudrun Goitein and Tony Lomax provided an update on proton radiation therapy of deep-seated lesions. Out of 99 patients treated on the gantry since 1996, 86 were still alive. "Local control was achieved in 72 out of 78 curatively irradiated patients and in 10 out of 20 palliative cases. Late toxicity greater than grade 2 was found in six out of 99 patients."

At the same time, there was no room for complacency. The Swiss Federal Office of Public Health's stipulation that good evidence was required if insurance coverage was to continue brought a new urgency to the collection of high-quality outcome data. The radiation oncologists at PSI, Damien Weber, Hans Peter Rutz and Beate Timmermann, under the supervision of Gudrun Goitein, got down to the task of constructing a database of every patient who had ever been treated on the gantry.

"We constructed it from scratch with an Excel sheet," says Damien Weber. *"We had to go through all the patients, give them an institutional number and then capture the basic patient characteristics, tumour characteristics, treatment characteristics – and then eventually, when that was done, seek out information on outcomes. We were phoning patients like crazy. This type of clinical research would not be possible today. But we were very motivated because we knew PSI was not sufficiently visible in terms of patient outcome and it was very important data to have. We needed to prove that proton therapy could be of value and to have figures and graphs to demonstrate it."*

The clinical research programme continued to grow.

Damien Weber: *"The PSI cyclotron underwent maintenance every Wednesday, which meant that you couldn't treat patients. So we had a whole day every week devoted to clinical research. Then, when the cyclotron shut down from 23 December until mid-April, all of our time could be dedicated to clinical research. So we had time to build our clinical research programme. We had a lot of encouragement from Gudrun, and it was important work, but we were also having a lot of fun doing it."*

Meanwhile, public awareness of proton therapy at PSI rocketed during 2002. In September, *Physics Today*, the flagship publication of the American Institute of Physics, published a special issue on "Physics Fighting Cancer" with an image of Gantry 1 on the front cover. Inside, Michael Goitein, Tony Lomax and Eros Pedroni contributed an article on treating cancer with protons. The introduction ran: "Once an obscure area of academic research, proton therapy is developing into an effective treatment option for use in hospitals."

As word spread among the public, a growing number of people desperate for a cure put themselves forward for treatment. An article in the Italian popular science magazine *Quark* entitled "Straight to the Target" presented PSI's Gantry 1 as a potential revolution in cancer treatment.

"The article brought a lot of hope to cancer patients in Italy willing to be treated in Switzerland," says Eros Pedroni. *"The central switchboard at PSI collapsed under the volume of calls in the days after publication. The calls of desperate patients brought tears at both ends of the line. In the end, the only solution was to prepare a pre-recorded response in Italian, communicating that the facility was currently shut down for winter."*

But if the public, government and sponsors were becoming increasingly excited about the potential of proton beam scanning, the same could still not be said of the scientific community.

In October 2003, as work continued on preparing the infrastructure for future installation of a new cyclotron and proton therapy facility at PSI, key staff from the programme travelled to San Francisco for the 39th meeting of PTCOG. At the time, Gudrun Goitein was the PTCOG chair. Damien Weber was among those presenting. But even before his session about the use of spot scanning for spinal tumours began, he was aware of discontented rumblings. Spot scanning was too complicated, people were saying. Passive scattering was a much better way to go. And at the end of his presentation there was only the politest of applause and barely a word of praise.

Chapter 7

2004–2008
A new lifeline for children

If Gantry 1 had proved the principle of proton therapy using pencil beam scanning, then Gantry 2 was designed to refine the technology. In the process it would demonstrate that pencil beam scanning was an effective approach that could feasibly be introduced into hospitals.

By the end of 2003, 166 patients had been treated for deep-seated tumours on Gantry 1 – 105 of them referred by Swiss centres and 61 from neighbouring European countries. The evidence was there that the technology was reliable and safe, reported Gudrun Goitein in the 2004 PSI Scientific Report.

In the same report, Goitein paid tribute to Hans Blattmann, who had retired in 2003. She reminded her colleagues that Blattmann was among the SIN physicists who had first published the concept of a proton therapy facility using scanning in 1989.

"Looking back on Hans Blattmann's professional career, we see that he was interested, and often engaged, in new developments and trends. Hans Blattmann is a physicist with broad knowledge and interests, and he had the flexibility to engage in new fields. That was certainly one of his many personal qualities."

However, PSI's experience at successfully treating patients using pencil beam scanning had made physicists and physicians alike aware of just how much could be improved. This needed to be addressed if the PROSCAN project was to bring higher patient volumes as planned. Foremost in their minds was the issue of how to overcome the problem of patient organ movement during scanning, which potentially jeopardised highly precise radiation targeting.

One way to overcome the problem was to scan the pencil beams so rapidly over the tumour that organ motion would become slow and insignificant in comparison. The tumour could be fast-scanned several times (or "repainted" with protons) to achieve the required dose.

"Gantry 2 was PSI's new platform for upgrading pencil beam scanning to perform at a speed close to the physical limits," says Eros Pedroni. *"The major drawback of the pencil beam scanning method is its inherent sensitivity to organ motion compared to the scattering foil technique. We envisaged solving this problem using much faster beam scanning techniques, the idea being to apply multiple target repainting at very high speed without compromising the size of the pencil beam. The end point of the development was the idea of treating the whole target volume of a repainting cycle in one go, within a breath-hold of the patient, in a time not longer than ten seconds."*

Though larger than Gantry 1, Gantry 2 would still be a "compact" gantry. It would re-use an important design feature of Gantry 1: the scanning magnets would be "upstream" of the main bending magnet – a far more technically challenging arrangement than conventional gantries, but one which made the gantry more manageable within a limited space.

The fact that the new gantry was being built alongside a completely new superconducting cyclotron and beam line meant that the entire system could be tailored for the kind of fast scanning now required. Rapid energy changes in the beam line could be achieved using a fast degrader system and laminated magnets. The beam intensity could be quickly modulated using a deflector plate at the cyclotron source. And a new system of magnetic parallel scanning would deflect the beam within milliseconds.

Whereas the magnets only deflected the proton beam in one direction on Gantry 1, with shifting in other dimensions achieved by patient table movement, the proton beam itself could now move in two directions, making the scanning process much quicker. The 210-degree rotation of the gantry, compared to 360-degree rotation of conventional gantries, would allow better access to the patient. It was a brave decision to limit gantry rotation, but one which presented no problems from a clinical point of view: the patient table could rotate fully in the horizontal plane, meaning there was no limitation in terms of beam delivery.

When David Meer arrived at PSI in October 2004, taking up the role of scientific officer in the Division of Radiation Medicine, Eros Pedroni handed him the recently finalised technical design report for Gantry 2. His first job was to build and operate a prototype of the scanning system in a dedicated test area, then produce the first dose distributions.

Meer, who had completed his PhD at the Institute for Particle Physics at ETH Zürich, had been taken on as part of the department's expansion plan. The team needed a physicist with expertise in electronics hardware and control systems. Christian Hilbes, a scientific officer in Eros Pedroni's team, knew the calibre of Meer's work from his own time taking a PhD at ETH. For his part, Meer had long known of PSI's work on radiation therapy. He knew that PSI's progressive programme was drawing some critical voices, but admired the institute's courage in investing so heavily in a second-generation gantry.

"In the mid-1990s, I remember that Gudrun Goitein came to ETH to give us a lecture on proton therapy, and it is one of the few copies of lectures that I still possess," he says. *"I thought it was an incredibly clever idea – a very useful application of the knowledge I had studied. I thought I could contribute something to this development. It felt like an honour to do so."*

Sharing an office with Hilbes, Meer became a central figure in the commissioning of Gantry 2 – a job that occupied him for the next ten years. Now head of technology development at PSI's Center for Proton Therapy, Meer looks back and regards bringing Gantry 2 into clinical operation as one of his greatest achievements.

"Eros Pedroni realised that you could make a machine as clever as you liked, but if you didn't think of the patient, it would fail. So when designing the gantry, he always thought of the patient point of view. His vision was that the patient should come into what looked like a normal-sized room, and all you saw of this massive machine was a half cylinder and a nozzle where the beam came out. It had to be comfortable, and there had to be good access to the patient with any treatment arrangement. This was the core idea of the whole mechanical design.

"Technically there were some big challenges," he says. *"Eros Pedroni was a visionary with some very clear ideas of how everything should work, and in principle everything was solvable. But in reality, if you have to make sure that the beam position is accurate to within 0.5 mm, it's easier to say than do. I remember that when we were in discussion with the engineering company about gantry construction, we explained the topology and upstream scanning to them. One of the managers told me that if he told his engineers to construct such a machine, they would say he is crazy, that it could not be done. That was two years before we got it all running. We managed to overcome all the challenges, and it's been running reliably ever since."*

Treating young children: beginnings

From the very start of PSI's experimentation with pencil beam scanning, it had been clear that one group in particular might benefit from such an accurate targeting of radiation: children with cancer. Because their bodies are still growing, their healthy cells are much more likely to be damaged by radiation, and they are far more vulnerable to side effects which might affect their lifespan or quality of life as they get older – for example, secondary cancer, learning disabilities, growth impairment and hearing loss.

For a child with a brain or spinal tumour, proton therapy offered both a highly concentrated radiation dose capable of neutralising a tumour and such accuracy that surrounding tissue was spared, making long-term adverse effects far less likely. Even those who criticised proton therapy on the grounds of cost had to acknowledge that, if pencil beam scanning reduced the likelihood of children becoming disabled and dependent, it more than justified the investment.

"We know that for children, treatment-related toxicity has a very high cost," says Damien Weber. *"We know that child cancer survivors in general are less fit, more sick, have lower-income jobs and are less well integrated into society than their peers. There are numerous cancer-survivor studies showing that. So that is an area where we are absolutely sure that delivering proton therapy provides good value for society, beyond saving lives."*

The cautious approach of Gudrun Goitein had meant that children under the age of 16 hadn't been treated on the gantry until 1999. By 2003, 16 children aged 7 to 16 had been treated, and there was strong evidence that proton therapy reduced the load of the radiation dose on normal tissue. With the benefits becoming clear, and the PROSCAN project providing the prospect of even more sophisticated scanning, Goitein was keen to get a programme for younger children under way. It would need to address the considerable challenges of treating very young children with cancer: children who could not be relied upon to keep still, and who would need to be anaesthetised or deeply sedated during radiation treatment.

In 2002 she had advertised for a new radiation oncologist at PSI, and at the interviews one applicant made it perfectly clear that she would like to establish a paediatric programme at the institute. Beate Timmermann, today the director of the Particle Therapy Clinic at the West German Proton Therapy Centre in Essen, had first heard about proton therapy using pencil beam scanning during her first job at the

University Hospital in Tübingen, where she specialised in radiation therapy for children. She knew this new type of particle therapy held immense potential for young children, and her enthusiasm impressed Gudrun Goitein.

When Timmermann got the job, she used the cyclotron's winter down time to research the subject further and made contact with the oncology and anaesthesia departments of the University Children's Hospital in Zürich. In Spring 2004, she approached Markus Weiss, head of the Anaesthesia Department, and Rita Feurer, an anaesthesia nurse, to discuss whether proton therapy for young children might be possible at PSI. What would the procedures be? What infrastructure would be needed?

"Anaesthesia isn't just required for young children because they are unable to lie still," explains Martina Frei, who was also part of the Zürich Children's Hospital team in 2004 and today leads the anaesthesia team for the paediatric programme at PSI. *"Even if you explain everything to a young child, they don't understand, and they are likely to be very afraid – alone in a high-tech room without their parents. At the time, there were only three proton therapy centres in the world where young children were treated under anaesthesia – Loma Linda, Boston and Tokyo – so we were really newcomers in the field."*

Weiss, Feurer and Timmermann put together their plans, and the children's hospital agreed that Weiss should create a paediatric anaesthesia service for PSI's Division of Radiation Medicine. It was the beginning of an important collaboration that continues to this day.

Children on the gantry
Having taken the decision to start treating young children, PSI began preparing an anaesthesia working area and recovery room with all the necessary equipment under the supervision of Markus Weiss and Rita Feurer. Staff were allocated and systems set up for the transportation of drugs and materials. Positioning procedures using bite blocks and body moulds were tested.

"From the beginning, our focus was on safety and high quality of care," says Markus Weiss. *"So we only used experienced paediatric anaesthesia personnel, and we equipped the anaesthesia and recovery room with all the material and technical support necessary for safe anaesthesia.*

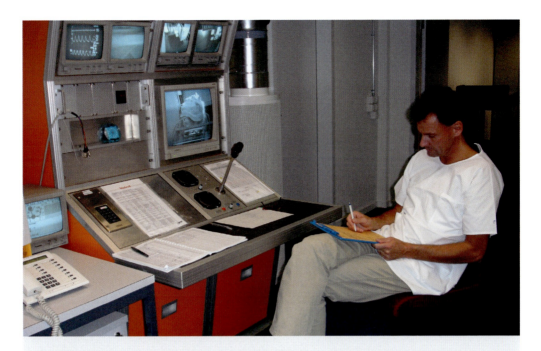
Figure 46. Markus Weiss supervising a child under treatment from the control room.

"In radiotherapy, the anaesthesia team is always outside the treatment room, so we can't just intervene if there is a problem. We have to anticipate possible problems before leaving the child alone in the treatment room.

"We have to be sure that the sedation is deep enough so that the child will not cough or move, but it should not be too deep, to prevent low blood pressure or airway and breathing complications. Other centres at the time were intubating the children under anaesthesia but we wanted to avoid this kind of airway manipulation because there might be adverse effects if a child was having treatments over many weeks. Many children with brain tumours require 30 fractions or more. So we provided deep sedation using an intravenous anaesthetic, propofol [diisopropylphenol], which was ideal for providing deep sedation but with the child still breathing on their own – or, as we call it, under 'spontaneous' breathing.

"During treatment, while the patient is alone in the gantry room, it is essential for the anaesthesia team to continuously monitor the patient via video. By wireless monitoring, we also check all the vital signs from the control room, using blood pressure, pulse oximetry, ECG, capnography and a microphone to relay heartbeat and breathing noise."

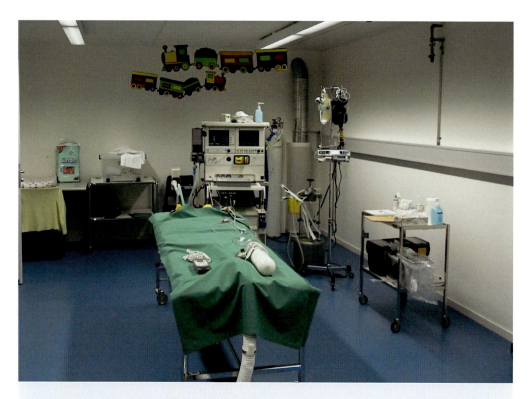

Figure 47. The original anaesthesia room, with anaesthesia machine, infusion pump, suction device, oxygen and pulse oximeter for transportation and materials trolley.

The first child patients to be treated under anaesthesia were selected not just by age but by weight. Children weighing more than 30 kg could not be treated on the gantry – not for medical reasons, but for safety. The limited access to Gantry 1's patient table meant it was challenging to remove heavier anaesthetised patients in the case of complications or fire in the treatment room. Children younger than one were at first excluded simply because of the higher risks of radiotherapy at that age.

The relatively remote location of PSI brought another stipulation, which continues to this day. With the closest hospital 16 km away, it was essential that the team treating children at PSI always had expertise to prevent adverse events and to cope if anything went wrong. It had to include a consultant medical doctor and a certified registered nurse, both specialised, trained and experienced in paediatric anaesthesia.

On 5 July 2004, the first child patient received proton therapy under anaesthesia on Gantry 1. He was just over two, and was treated for a rhabdomyosarcoma in the orbit of his eye. Seven days later, the second child, aged just one year and eight months, started treatment for a brain tumour.

In September, a child received a treatment in the prone position for the first time – after the team had addressed significant technical challenges. The boy had a type of brain tumour called a medulloblastoma.

"The boy eventually needed an intubation for all his 32 treatments, because under deep sedation he either had an airway obstruction and was not breathing sufficiently, or he was breathing but moving," says Martina Frei. *"It was a big challenge. So we intubated the child for all 32 fractions, even though we were worried that it might cause long-term damage to his larynx and trachea. But altogether it went very well. We were able to see him again at a follow-up some years later at the children's hospital in Zürich, and we did a rigid endoscope airway examination to see if there were any lasting effects from our daily intubation. And we were very happy and proud that we could see no sign of any damage in the trachea.*

Figure 48. A child being prepared for treatment on Gantry 1. Monitoring and anaesthesia equipment can be seen at the front of the patient table. Picture: Paul Scherrer Institute

"Later we found out how to place the children in a better way to keep the airway open in the prone position."

By the end of 2005, 20 young patients had been treated under anaesthesia. They are among the first children in the world to have been treated with proton therapy using pencil beam scanning while sedated with spontaneous breathing. Despite early worries about the effect of daily anaesthesia repeated over several weeks, the evidence was that children tolerated this well, with no major side effects. Concerns about the effect of long-term tolerance to propofol were also dissipated.

Martina Frei acknowledges that there were stresses in the early days – some caused by technical problems, meaning delays in daily treatment schedules. But she always knew that she was engaged in a significant project. And the safe and successful treatment of children brought real highs. Radiation therapy for children at PSI could be – and still is – very emotionally involving.

"Anaesthesia is always risky, particularly in smaller children," she says. *"The special risk with radiation therapy is that you are at a big distance from the child – you*

Figure 49. Martina Frei welcomes a young patient and her mother for treatment preparation.
Picture: Paul Scherrer Institute

can't just press the infusion pump and give a little more anaesthetic if the child starts to come round. Everything has to be planned for. You can't break or postpone radiation therapy if a child has a cold or feels unwell, so you have to prepare for that too, preparing them with suctioning and drops in the anaesthesia room so that you can be confident that they can breathe properly. For very ill children, it's a matter of balancing risk versus benefit.

"Working in the PSI team can be much more emotional than much of our work at the children's hospital. As an anaesthetist in hospital, we don't treat the same patient for a long period and our contacts are generally very short. Most patients we see just for the operation or intervention, and then in the recovery room before they go home or to the ward. But here at PSI we see these children for six weeks or longer, and we build up an emotional connection with them and their family. This makes it special.

"What I love about work at PSI is that it's not just technical. Caring for a child, giving support to parents when they are completely down – all these things make it really special. Very often we see children arriving anxious or traumatised by the treatments they have had before – sometimes crying when they see someone in a white coat. And it can be very intimidating to come to a big technical place like PSI for treatment. But over the years we have made the anaesthesia room as friendly and cosy as possible. And up until now we have managed it so that all of our children, in the end, come for treatment with a smile. They like to come to us."

Evidence of the success of proton therapy at PSI using pencil beam scanning continued to grow. During 2004 and 2005, Damien Weber and his collaborators at PSI published evidence showing that the increasing number of adult patients being treated for chordoma and chondrosarcoma at the skull base and along the spinal cord were doing very well: depending on the tumour site, there was a three-year survival of 89 to 93 per cent.

Then came a pause. PSI's entire proton treatment programme – including the treatment of children under anaesthesia – had to stop for the whole of 2006 so that the complicated process of connecting all proton therapy facilities to the new dedicated medical cyclotron could begin: first Gantry 1, then the new OPTIS facility and ultimately Gantry 2.

The test area

In 2004, the COMET superconducting 250 MeV cyclotron, designed to provide stable beams to the gantries and OPTIS throughout the year, had finally been delivered and installed in the area designated for a new proton therapy facility. The first proton beam was extracted on 1 April 2005. David Meer still has a paperweight – a metal replica of COMET – commemorating the landmark.

A test area was set up, with a beam line running to the area where construction of Gantry 2 would soon begin, and David Meer was tasked with evaluating the performance of the beam line, dosimetry equipment and control systems.

"The COMET superconducting cyclotron that ACCEL developed with us turned out to be a great success story, and was the model for many others sold around the world," says David Meer. *"It is called C1, because it was the first, and the knowledge we generated from it was incorporated into subsequent machines that have been installed all around the world.*

"In the test area, we had the chance to test the two-dimensional scanning and develop the control systems. I was working with Christian Hilbes, a physicist with a high energy physics background, and he took over responsibility for the control system architecture.

"At the time, in 2005, Gantry 1 still had the beam from the large ring cyclotron, so it was very comfortable for me in the test area. I always arrived at eight in the morning, went to the test area, asked for the beam and had it for as long as I wanted. I did not appreciate it at the time – today, if I want to do an experiment, I have to go

Figure 50. The commemorative COMET paperweight. Picture: Paul Scherrer Institute

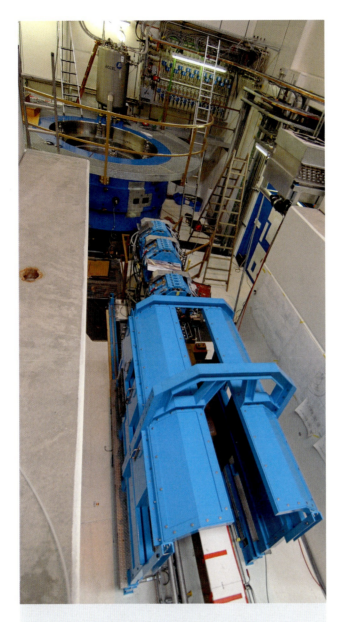

Figure 51. Installation of the COMET cyclotron in 2004. In the foreground is the degrader, to vary energy in the pencil beam. Behind it on the beam line is the kicker magnet, to switch the beam on and off during scanning. At the rear left is the cyclotron. Picture: Paul Scherrer Institute

in at the weekend or start at eight in the evening! But, you see, from 2005 to 2007 there was a lot of development to be done until we could connect Gantry 1 to the new COMET superconducting cyclotron. And during these two years, PSI solved many issues related to connecting the existing technology to a new source of protons.

"As it was the first machine of its kind, we had a steep learning curve. The idea was always that the dedicated cyclotron would have reliability, and indeed for ten years we never had a longer break than three or four days. Nowadays, I have to say it's really a smooth machine. We have a reliable and stable beam current."

The suspension of patient treatment while Gantry 1 was connected to the new cyclotron was supposed to be brief, but in the end it took a whole year.

"When we started, we told people it would take six months," says Martin Jermann. *"But you can't hurry, and you have to complete all the checks to ensure that it is safe to start using on real patients. We had new safety systems now, with the new cyclotron, and they all had to be tested thoroughly."*

Figure 52. The COMET cyclotron being installed, with the cover not yet in place. The proton source is at the centre. Protons accelerated by rapidly changing high voltage are kept on a circular orbit by the spirally shaped surface of the magnet. Picture: Paul Scherrer Institute

Gantry 2 installation

The large components for Gantry 2, including its massive 90-degree bending magnet, began to be installed in 2006. The control and safety systems used on Gantry 1 had to be adapted for it – so the team of software and hardware developers needed to be enlarged.

"Christian Hilbes brought in many new ideas for software and hardware development, but left PSI after a few years," says Martin Grossmann. *"Christian Bula and Alex Mayor, who had been developing industrial software, joined the controls team. It was decided to stay with the platform used in Gantry 1 but to rewrite all software and to separate the core application from the graphical user interface. This new version of PSI's control system is still used both in Gantry 2 and in OPTIS.*

"We were successful because we had a small team, with expert support from other departments at PSI, which allowed us to be very focused. We discussed risks and mitigation measures together over many hours, with people from different backgrounds – medical doctors, physicists, engineers – then writing documents and reviewing them."

Eros Pedroni says that, although slow, the construction of Gantry 2 was relatively problem-free – apart from one major hurdle. *"Gantry 2 was realised in one go, without the stringent compromises that we had to make with Gantry 1 due to cost constraints. The only delay effect was the priority given to the realisation of the new OPTIS2 facility.*

"However, the realisation of the huge last 90-degree bending magnet was a major challenge. It required a very sophisticated design of lamination so that it could provide very fast two-dimensional parallel scanning at the isocentre. The lamination was planned to be vacuum-tight but the industrial provider couldn't achieve this specification. The magnet group of PSI, led by David George, bravely re-machined the magnet in a local Swiss workshop and brought the system to its final perfection. Thus they saved the project."

Figure 53. The main mechanical component of Gantry 2 being inserted through an opening in the hall roof.
Picture: Paul Scherrer Institute

Members of the international particle therapy community had a chance to see the Gantry 2 site at first hand in June 2006, when the PTCOG meeting was held for the third time in Villigen. This was an excellent opportunity for PSI to promote its proton beam scanning technology.

At the meeting, David Meer described the next-generation control system which would "open the door to explore the advanced scanning modes of the new Gantry 2". Dölf Coray described the extensive quality-assurance processes being put in place on the gantry. Christian Hilbes explained how safety systems were being upgraded into central and local subsystems under PROSCAN. And Gudrun Goitein looked back on nine years of using spot scanning technology at PSI.

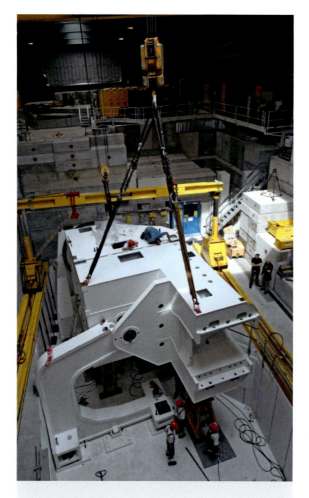

Figure 54. The component is then mounted on the floor of what will become the treatment room.
Picture: Paul Scherrer Institute

"Introducing a new technology requires great care concerning the safety and reliability of the entire system," she said in her report. "Treatment outcomes are analysed not just with regard to local control and survival, but also with careful attention to toxicity and possible unexpected effects. Apart from favourable outcomes, we also learned about limitations – which are inherent in radiotherapy as well as to the use of scanned proton beams."

The team had focused, she said, on issues affecting dosimetry such as organ and target motion, with the intensity modulation techniques devised by Tony Lomax playing a major role. In addition, a great deal of attention had been paid to optimal patient immobilisation.

"Outcome results are very satisfying," said Goitein in her report. "The introduction of a dedicated medical accelerator, and the next generation proton gantry, will allow us to overcome existing limitations to a large extent, and to further improve spot scanning radiotherapy with protons."

The OPTIS2 facility

The installation of the new OPTIS facility within the new proton therapy facility began in the second half of 2007 and continued through to late 2009. Unlike the gantries, where the beam moved around the patient, the new OPTIS facility contin-

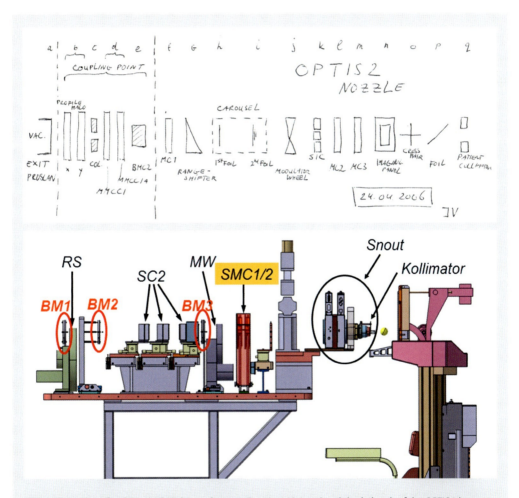

Figure 55. Designs for the new OPTIS2 equipment. Top: Jorn Verwey's original sketch of the OPTIS2 layout, dated 24 April 2006. Below: a more complete diagram of the arrangement. It shows: beam monitors (BM), ionisation chambers measuring proton beam intensity; range shifters (RS), which adapt the proton energy to the clinical requirement; scattering foils (SC), which broaden the beam to cover the tumour; a modulator wheel (MW); segmented monitor chambers (SMC), which measure the proton beam profile and intensity; the nozzle *(snout)* and collimator *(kollimator)*.

ued to use a fixed horizontal proton beam, with the patient positioned in front of the beam and having to gaze continuously at a fixed light point for up to a minute.

The 250 MeV proton beam from the new medical cyclotron needed to be degraded to 70 MeV to provide the right energy for eye treatment. However, in the degradation process, 99 per cent of the protons extracted from the cyclotron would be lost, meaning that beam intensity would be less than with Injector 1, previously used for eye treatments. This meant that a completely new proton delivery system which would bring the beam up to sufficiently high intensity had to be built for OPTIS2. Jorn Verwey, the medical physicist who succeeded Emmanuel Egger as head of OPTIS, led development of the new system. Marco Schippers led work on tuning the beam line.

As Verwey wrote in a 2010 paper describing this second-generation beam line: "The technical challenge was to increase beam line transmission without significantly altering existing beam characteristics on which the clinical success is based." [20]

To keep treatment times to a minimum, a "double-scattering" process was to be used within the OPTIS2 nozzle, to ensure a sufficiently high proton beam intensity. A range shifter would accurately set the desired range and then a scatter foil would broaden the beam and ensure a homogeneous field. Two X-ray units would be used to exactly align the patient's eye with the treatment isocentre.

"The comfort of the patients was also improved," says Leonidas Zografos. *"The new facility included a stereotactic chair with electronic motion, a computer-based control room and a digital imaging device."*

Treatment resumes
After the year-long shutdown for connection to COMET, patient treatments on Gantry 1 resumed in February 2007 – using, for the first time, a dedicated medical cyclotron. There was another brief shutdown between June and mid-August, when

---

[20] J. Verwey, F. Assenmacher, J. Heufelder, M. van Goethem, M. Grossmann, G. Goitein, T. Lomax, A. Tourovsky, L. Zografos, E. Hug, 'OPTIS2: A Second Generation Horizontal Beam Line for Ocular Proton Therapy using a 250 MeV Cyclotron – First Patients Treated 25 Years after Start of the OPTIS Program at Paul Scherrer Institute', *International Journal of Radiation Oncology, Biology, Physics*, 78:3 Suppl. (2010), S132–33,

Gantry 2 and the new OPTIS facility were connected to the new beam line. And then began a new era of regular treatments at a dedicated proton therapy centre.

By this time, Eugen Hug had become the director of the Center for Proton Therapy. As he noted in the PSI Scientific Report for 2007: "The number of treatments per day has been ramped up steadily to about 15 patients in a continuous five days/week treatment program. For the first time at PSI, patients were treated between Christmas and New Year, demonstrating the autonomy of the systems and infrastructure."

There were, however, tensions as the drive for technological excellence in the emerging Gantry 2 temporarily competed with patient care on Gantry 1. These were acknowledged in the 2007 scientific report.

"Patients are now usually treated Monday through to Friday, one treatment per day, five times per week. Treatment schedules are therefore similar to conventional radiation oncology centers. We try to balance patient treatment demands with the requirements of maintenance, technical improvements and innovations, and commissioning of the new OPTIS2 and Gantry 2 treatment areas. Balancing competing demands for beam time, we currently arrive at a patient volume for deep-seated tumors on Gantry 1 of 15 patients per day on a regular and ongoing basis as well as taking care of the eye program's 200 to 230 annual patients."

The balancing act wasn't always easy. According to Martin Jermann, after the delays in installing the new cyclotron, the need to keep up patient flow contributed to delays in commissioning Gantry 2.

"After the shutdown in 2006, the clinicians said they would like the full year to treat patients – they had waiting lists which needed to be a priority. We also had to find time to test Gantry 2, and I remember Eros Pedroni coming to me saying he needed more people to be able to do this. Tony Lomax, for example, was needed to implement the treatment planning system on Gantry 2, but he was also aware of his responsibilities to work with clinicians on treating patients on the waiting list. So there was a certain conflict."

Another source of delay were safety regulations. As with Gantry 1, PSI had to convince the Swiss Federal Office of Public Health that they should have an operating licence to treat patients on Gantry 2. There were now many projects developing proton therapy worldwide, and new international safety standards were being set that had to be complied with.

"It took maybe one and a half years for us to get approval to treat patients in our new facility," says Martin Jermann.

But there was an impressive, and growing, body of evidence that the massive effort being invested in cutting-edge radiation therapy at PSI was more than worth it.

The evidence of success
The 2007 PSI Scientific Report records the long-term clinical results of proton therapy at PSI. The eye programme had treated more than 4,800 patients and achieved local cancer control in 98 per cent of them. Treatment for deep-seated tumours with scanning technology now represented "the world's leading technology for proton delivery". By the end of 2007, 315 patients had been treated with pencil beam scanning at PSI. Long-term success rates were available for the first time, with five-year local control rates for chordomas and chondrosarcomas of the skull base treated between 1998 and 2005 standing at 81 per cent and 94 per cent respectively.

Evidence also indicated that these results were better than with conventional radiation treatment. "First and foremost it means that patients with skull base tumors previously considered fatal, now have a chance of being cured," asserts the annual report.

For those working at PSI, another highly significant benchmark had been reached. The results were comparable with proton therapy at Loma Linda University and Massachusetts General Hospital – the gold standard that Gudrun Goitein had specified PSI had to reach back in 1996 when treatment with pencil beam scanning began. "PSI results appear to be at least equivalent to those obtained at world-renowned cancer centers, for example at the Massachusetts General Hospital, USA," said the report.

Now PSI's success in treating previously untreatable cancers was beginning to attract support. In 2008, the Swiss Canton of Aargau – in which PSI is located – decided to support the construction of Gantry 2 with a Swiss Lottery Fund allocation of 20 million Swiss francs. Meanwhile, the Swiss Cancer League granted financial support to the child proton therapy programme at PSI. At the same time, multidisciplinary treatment protocols for child malignancies in Europe began to include proton radiation therapy in their treatment guidelines.

Beate Timmerman was all too aware of proton therapy's enormous potential. Authoring a new study on the use of pencil beam scanning for child soft tissue sarcomas at PSI, she reported preliminary data suggesting "excellent acute tolerance" and "satisfying results" after a short observation time.[21]

"At the Paul Scherrer Institute the expansion of the proton facility is ongoing," she wrote. "Paediatric patients will be preferred, as they are considered the individuals potentially benefiting most from proton therapy."

Smiling water and star dust
As the only centre in Europe providing proton therapy to young children under anaesthesia, PSI was, by the late 2000s, receiving young patients from all over the globe: Germany, the United Kingdom, France, Italy, Spain, Scandinavian countries, the Netherlands, Belgium, Russia, South Africa, China and Singapore.

Procedures to treat children on Gantry 1 continued to develop. For example, a new patient transporter made it easier to adapt the height of the couch and position the sedated child patient in the anaesthesia room before going to the CT room. Care facilities for the children and their families also improved, forming the basis of the highly personalised support services that are in place today.

"The families are in an extreme situation when they bring their child for proton therapy," says Martina Frei. *"They are under stress because of their child's diagnosis itself, and the child is often in a deteriorated condition and traumatised by the treatment they have already received. Then they have to bring the child in for a daily treatment for about five or six weeks, sometimes even more. The child has to be fastened, sedated and treated each day, which can take many hours – especially if you take into account transport to and from PSI, waiting for treatment, delays because of technical problems, time for the child to awake afterwards.*

---

[21] B. Timmermann, A. Schuck, F. Niggli, M. Weiss, A. J. Lomax, E. Pedroni, A. Coray, M. Jermann, H. P. Rutz, G. Goitein, 'Spot-scanning proton therapy for malignant soft tissue tumors in childhood: First experiences at the Paul Scherrer Institute', *International Journal of Radiation Oncology, Biology, Physics,* 67:2 (2007),: 497–504. doi: 10.1016/j.ijrobp.2006.08.053.

Figure 56. The anaesthetist accompanies a deeply sedated patient on the transportable couch to Gantry 1, along with equipment needed during treatment. Picture: Paul Scherrer Institute

"Maybe the parents are trying to organise their work around all this. Maybe there are siblings of the sick child who need support. Families from abroad have special problems: they are far away from their social network, and often don't speak German or English. Sometimes they have to bring the whole family to Switzerland with them.

"Our aim has always been to create an atmosphere in which all the family can feel comfortable and safe – a kind of oasis where they can talk about their fears and hopes and where they can find comfort and support. We decorate the rooms. We offer the children toys and books. We distract them with fantasy stories and finger puppets. We have given names to the anaesthetic drugs we use: we have 'spaghetti water' for NaCl 0.9 per cent, 'smiling water' for midazolam, 'elephant milk' for propofol and 'star dust' for ketamine or nalbuphine.

"The parents are allowed to accompany their child until he or she is sleeping. Even siblings, grandparents, uncles and aunts are allowed to accompany the child to the anaesthesia room. Sometimes siblings hold the hand of the sick child or sing a song until he or she is asleep. This is very special.

Figure 57. Rewards available for child patients after irradiation include small beads, kept in drawers, and finger puppets. Picture: Paul Scherrer Institute

"We are happy and proud that, up until now, all the children we have treated have lost their fear and liked to come to us, even if they were in a panic at the beginning."

The world starts to listen

As the PROSCAN project began to take shape, and PSI stood on the brink of being able to treat large numbers of patients with cancers that others would regard as untreatable, the decisions the PSI directorate had taken looked increasingly justified. Martin Jermann and Gudrun Goitein had never doubted that the massive investment in a new kind of proton therapy facility would be worthwhile. But now there were signs that the wider world was beginning to share their confidence. Other clinical facilities were investigating the possibility of using proton beam therapy with pencil beam scanning rather than scattering techniques.

Jermann remembers that Zelig Tochner, Professor of Radiation Oncology at the University of Pennsylvania (Upenn) School of Medicine, visited PSI several times from 2000 onwards and was very excited about PSI's gantry projects. He took what he had learned about scanning technology home to the United States.

"He was one of the first in the US to push in this direction. However, Upenn commissioned a proton therapy facility that used passive scattering and double scattering, which opened in 2010 and operated for several years before they decided to upgrade their treatment rooms to pencil beam scanning technology. It was a similar story at the proton therapy centre at the MD Anderson Cancer Center in Texas, which opened in 2006. It was designed to use passive scattering, but before they went into operation, they decided that one of their gantries should use scanning technology." Their first scanning gantry came into operation in 2008.

"Centres started promoting scanning technology. And an important reason for this, which only became clear in an American commercial hospital environment rather than a research institution, was cost. Their major argument was that you could have a much higher throughput of patients through the treatment room with scanning technology.

"With passive scattering, you need to replace the collimators for each field to conform the beam to the tumour shape. After the first field, you have to stop the treat-

Figure 58. An aerial photo of PSI in 2009, showing the Center for Proton Therapy to the left of the crane. The "medical pavilion" would soon be extended into a full medical facility building.
Picture: Paul Scherrer Institute

ment, go into the treatment room, replace the collimator, go out of the room, make all the tests, then treat with the second field, and so on with maybe four fields. So you lose a lot of time compared with scanning, where you do it all in one go.

"When MD Anderson started using scanning in their third treatment room, they realised that they could treat as many patients per day in this one room as in the two others."

The irony was that, after all the criticisms endured by PSI about the high expense and complexity of scanning technology, many were coming around to it on the grounds of cost.

The first scanning gantry in Europe outside of PSI came into operation in 2009 at the Rinecker Proton Therapy Center in Munich, Germany. But Martin Jermann knew that, despite PROSCAN taking longer than expected, PSI was still ahead of the game.

The scattering vs scanning debate may not have been won yet, but by the end of the 2000s the balance of opinion was changing rapidly. Looking back, Damien Weber thinks that from the clinical point of view of radiation oncologists, there was only ever one winner.

"Today, pencil beam scanning rules. One of its major advantages over passive scattering is that you completely control the deposition of dose, the end point of the Bragg peak, throughout the tumour. So the conformality of the dose is way better than scattering, where you have to position the dose peak on the posterior of the tumour. The integral dose is also better with pencil beam scanning."

At the 2009 PTCOG meeting at the German Cancer Research Center in Heidelberg, Germany, there were strong signs that many were beginning to think the same way. Jay Flanz from the Massachusetts General Hospital presented a paper on "From Passive to Active Beam Delivery at MGH". Eros Pedroni was there too, with a PSI contingent, and gave a presentation on the second-generation scanning proton gantry at PSI. Six years on from the PTCOG meeting in San Francisco, when Damien Weber's presentation on pencil beam scanning was met with barely disguised disdain, the reception given to Pedroni only served to emphasise how radically attitudes were changing.

"The Congress centre room used for the physics session was not large, but there were chairs for at least 50 to 80 people," recalls Martin Jermann. "And when it came to Eros's lecture on scanning technology, I saw something I had never seen before. It was the only session at the whole meeting where there was not enough room. Everybody was very interested to know what was going on in the field of scanning. The room was so full that people were sitting on the floor at Eros Pedroni's feet."

Chapter 8

2009–2021
The new era of proton therapy

Beneath the pragmatic description of proton therapy activity and results, the 2009 PSI Scientific Report reveals more than a faint glow of pride. "Expansion based on continuity: technology projects in proton therapy justified by clinical results," ran the heading. All that innovation and investment, from the earliest days of pion therapy in the 1980s to the state-of-the-art Gantry 2, was demonstrating its value.

"As enthusiasm for spot-scanning-based particle therapy continues worldwide," read the report, "and as other centres implement first generation systems, the issue of the 'safety and efficacy' of spot-scanning-based proton therapy, compared with the historically used passive scattering technology, becomes of paramount importance. Only PSI is presently able to provide these important clinical data, with 517 patients having been treated on Gantry 1 from 1996 to October 2009 . . .

"Our data provide the medical evidence for the 'safety and feasibility' of spot-scanning technology for the clinical indications presently treated at PSI [mainly difficult-to-treat tumours of at the skull base and next to the spinal cord, and tumours in infants and children]. The research effort at PSI in the field of particle therapy of deep-seated tumours, starting with pion therapy in the 1980s, has now come full circle.

"From initial design and treatment concepts, to early research, manufacturing and clinical implementation, to ultimately routine use and now proof of not only principal, but actual, readiness for widespread clinical implementation, is an outstanding accomplishment by literally one generation of researchers at PSI.

"The ability to control tumours, and essentially cure patients in many cases in which an actual cure was previously rarely an option, is an extremely satisfying result."

The authors of the report were Eugen Hug (acting head of the Center for Proton Therapy), Gudrun Goitein (who had stepped down as head two years earlier),

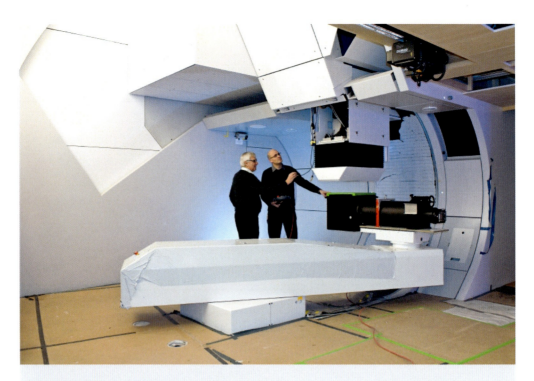

Figure 59. Eros Pedroni (left) and David Meer in discussion during the commissioning of Gantry 2.
Picture: Paul Scherrer Institute

Christian Bula (responsible for control systems), David Meer, Eros Pedroni and Jorn Verwey – two of whom (Goitein and Pedroni) had been working on the pion therapy project two decades earlier.

This didn't mean that the team felt there was any room for complacency – or need to slow the pace of development. "We feel therefore encouraged to proceed with the next generation project of spot scanning, to be realized on the new Gantry 2 system," said the authors.

Despite the delays caused by prioritising the OPTIS2 start-up and keeping the clinical programme running on Gantry 1, Gantry 2 was now taking shape. Tests had proved that it was possible to paint the dose on a tumour with the proton beam moving at very high speed. The two fast magnets controlling the beam movement in two dimensions moved at speeds of one to two centimetres per millisecond. In the third dimension, the depth of proton penetration into the tumour could be changed from one tumour layer to the next in less than 100 milliseconds through adjusting the protons' energy levels. This made "volumetric repainting" possible, with the same volume being painted with radiation several times in a single treat-

Figure 60. Installation of the CT imaging system (right), within reach of the patient table (left).
Picture: Paul Scherrer Institute

ment session. Shaping of the dose could be further refined by changing the beam intensity at the cyclotron source.

Two other important innovations were being incorporated. One was a sliding CT imaging system in the treatment area, within reach of the patient table. This would allow the patient to be positioned highly accurately using the most advanced imaging technology available. Previously this had been done in a separate CT room and the patient then had to be pushed into the treatment room.

The other development was a "beam's eye view" X-ray system. The layout of Gantry 2, with double parallel scanning occurring "upstream" of the 90-degree bending magnet, made it possible to take X-ray images in the beam direction simultaneously with the proton beam. This feature would not only help with patient positioning, but "open the door for developing new and more powerful quality assurance tools for controlling target motion during proton beam delivery," said the PSI Scientific Report.

The first beam line was successfully sent through Gantry 2 on 9 May 2008.

"We invented many different things ad hoc as we progressed, and also during installation," says David Meer. *"This kept things flexible, but it's probably not something we could have done in a commercial environment."*

PROSCAN support
The PROSCAN project was supported by the following sponsors and donors*:

    Swisslos-Fonds Kanton Aargau
    Swisslos-Fonds Kanton Zürich
    Swiss Cancer Research
    Walter Haefner Foundation, Zürich
    Cancer League Kanton Zürich
    Genossenschaft zum Baugarten, Zürich
    A. Meier-Schenk
    Dr Leopold and Carmen Ellinger Foundation
    Kanton Aargau
    Dr P. Alther, Zollikon
    Jubilee Foundation CS Group, Zürich
    Hilti Family Foundation, Schaan
    Hugo and Elsa Isler Fonds
    Pfeiffer Vaccum AG, Zürich
    VAT Vakuumventile AG, Haag
    R. + S. Braginsky Foundation, Zürich
    Thalmann Foundation, Olten
    Swiss Foundation for Clinical Cancer Research
    Degler GmbH, Weinstadt-Beutelsbach
    Cancer League Kanton Aargau
    Berthold Technologies GmbH
    Dres B.&J. Rust Good
    Busch AG, Magden
    E. Stehli
    Migros-Genossenschafts-Bund, Zürich

*Donations of more than 5,000 Swiss francs

OPTIS reborn

Twenty-five years on from its launch as the first ophthalmological proton therapy unit in Europe, OPTIS had treated 5,300 patients. Now it was about to be reborn. During 2009, the foils required for the OPTIS2 double-scattering system, the digital imaging system, patient-positioning robotics and software were all installed. Testing showed that the new OPTIS beam characteristics almost matched those from the Injector 1 cyclotron, and beam symmetry was significantly improved.

By June 2009, control and safety systems were all complete. OPTIS2 was audited by the Swiss Federal Office of Public Health and a treatment permit was obtained after completion of a quality-assurance programme. The first OPTIS2 patient was successfully treated in January 2010 and by the end of the year, 47 patients had been treated on OPTIS2.

This meant that the old Injector 1 cyclotron could finally be shut down. *"This was a relief,"* says David Meer. *"For a few years, Injector 1 had only been running for OPTIS, and its operation had become more and more demanding, with leaks in the cooling circuit having to be regularly repaired."*

Figure 61. The official opening of OPTIS2 in 2010. Eugen Hug cuts the ribbon with Martin Jermann (right). Picture: Paul Scherrer Institute

Ann Schalenbourg, today the head of the Adult Ocular Oncology Unit at the Jules Gonin Eye Hospital in Lausanne, had joined Leonidas Zografos on PSI's OPTIS programme in 1995. The OPTIS2 upgrade was highly significant for the continuation of the eye therapy programme, she says.

"At the very beginning of OPTIS in the early 1980s, it took Charles Perret months to build a chair," she says. *"But now we had gone from the technical era to the computer era and everything needed to be reinvented. It took the team three years to reinvent OPTIS."*

The redevelopment allowed PSI's eye programme to function far more effectively – integrated into a dedicated clinical facility where patient treatment was a priority rather than something squeezed in between experiments. Despite the new technology, the treatment itself had not fundamentally changed – but it was now interleaved with gantry treatments, making it possible to operate continuously throughout the year rather than in once-a-month batches.

This had significant implications for how the Jules Gonin collaborators worked. In the days of the old OPTIS facility, Ann Schalenbourg had made visits to PSI one week every month because that was when the Injector 1 cyclotron was made available to them.

"The Jules Gonin Eye Hospital has always taken the responsibility for getting patient referrals, diagnosis, management and follow-up – all the clinical side, really – with PSI providing the technical delivery of protons. Initially there was an agreement that there needed to be a responsible ophthalmologist present during treatment, so before 2010 we went to PSI physically during the week we had allocated to us for treatment. First it was Zografos for two days and me for two days, then gradually that evolved so it was just me at PSI, one day a month.

"With OPTIS2, patients could be treated weekly rather than once a month. We had a maximum of eight slots every day of the week, but it became impossible for us to travel to PSI every week, three hours there, three hours back. So we stopped travelling to Villigen and today have a weekly online video meeting to discuss treatment plans with all the clinicians and physicians involved. The PSI radiation oncologist, Alessia Pica, visits us weekly to study patient files, do follow-up and participate in clinical research."

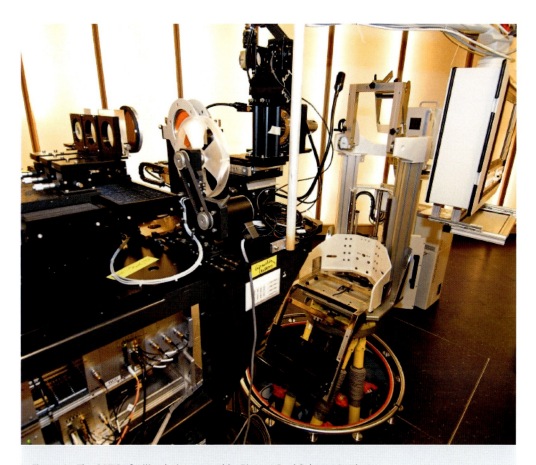
Figure 62. The OPTIS2 facility during assembly. Picture: Paul Scherrer Institute

Today, the logistical challenges that exist are largely the result of the number of referrals that come from all over Europe, based on the facility's long-standing reputation. Ann Schalenbourg is full of admiration for her team at Jules Gonin who help newly diagnosed patients through their post-diagnosis panic, and help them plan for treatment at PSI.

"Many patients have just discovered they have an eye tumour when they contact us. Depending on the tumour's size and the degree of retinal detachment, there might be only a couple of months to save the eye. So they need to get through the language barriers, the administration barriers, sort out their family, organise travel to Lausanne, organise getting to PSI for treatment. All the while they have to cope emotionally with the fact that they have a malignant tumour in their eye, and that they could die of metastases. Our secretaries play an important role in helping them through all that.

"We work hard to maintain our reputation and the confidence of everyone we work with, constantly re-explaining to all those involved what is going on. A small mistake might have significant implications and worsen prognosis."

OPTIS achievement

Ann Schalenbourg sees evidence of the success of the OPTIS programme in the experiences of patients she has treated: a professional photographer, a dancer, a jazz musician and many others from all walks of life whose lives and livelihoods have been saved by proton therapy and who will be followed up for years after treatment. She remembers treating a patient of 98, who had already lost sight in one eye, for whom saving his remaining sight meant the difference between independence and becoming a burden. She saw him at a check-up three years later, and he was still living in his own home.

At the other end of the scale, she remembers a child who was treated for retinoblastoma at the age of four, under the supervision of the paediatric ocular oncologist at the Jules Gonin Hospital.

"At Lausanne we have the reputation for saving eyes that others would not. We have separate units for adults and children, and generally younger children do not get treated at OPTIS because they cannot collaborate actively. But this was an exceptional child who was four years old but very mature for his age, and he had a recurrence of retinoblastoma. The only way to save his eye was through OPTIS treatment.

"So my colleagues and his father had to think of ways of keeping him calm, wearing the positioning mask and keeping him looking at the fixation light during treatment. They made a story out of it: he was the knight who had to combat the evil by accepting the mask and looking at the light. He went along with this, at the age of four, because it made a hero out of him. And every time he came out of a treatment session he got an ice cream. I can tell you we didn't have any problems with any of the adult patients who were waiting, after they saw a four-year-old coming out!"

But the success story of OPTIS and then OPTIS2 goes beyond the anecdotal. The database first set up by Emmanuel Egger in the late 1980s has stood the programme in good stead and provided authoritative evidence of the high quality of the treatment and its long-term effectiveness.

"The common database contains data on every patient we have treated at PSI since the mid-1980s and it is now maintained and co-financed by both institutions, PSI and the Jules Gonin Eye Hospital," says Ann Schalenbourg. *"We continually produce scientific papers reporting on the results in the database – which is our duty as a university institution, and is a means of quality control."*

The research shows that for more than 90 per cent of patients treated to date, it was possible to save the eye. In 2009, PSI's Scientific Report included the results of two large outcome analyses which showed that if local eye tumour control was successful, it had a significant effect on survival. For ocular melanoma, local control rates of 97 per cent were being achieved at five years, providing "proof of principle" for the high-dose, hypo-fractionated irradiation provided at PSI.

By the end of 2021, 7,800 patients had been successfully treated for ocular tumours at PSI since 1984. More than 1,800 patients have been treated at the OPTIS2 facility. Ann Schalenbourg says that no other proton therapy facility in the world has treated so many eye patients, or treated them so well.

"There are now around 15 other centres like ours in the world, but we have reported the best results. Ever since 1988 we've been looking at the reasons for local recurrence, and trying to address them. So while most proton centres are pleased with a 94 to 96 per cent local tumour control, we have reached a 98.8 per cent control at five years."

A convergence of many factors account for this excellence, says Schalenbourg.

"I think it's largely the long learning curve we have, the inheritance of Professor Zografos and collaborators at PSI going back to 1984. Zografos's brilliance and his inability to accept failure was very influential. Other centres let junior doctors do the surgical procedures, such as putting on the tantalum clips, before radiation therapy. This is brilliant for training, whereas our approach, allowing only experienced surgeons, is brilliant for patients. We also have the advantage that we make personalised collimators tailored for every patient, whereas other centres use three standardised collimators. There's something of Swiss cultural perfectionism in that.

"I've felt extremely privileged to be part of this, and as a doctor it is very special to be able to offer patients what you are convinced is the best treatment in the world.

"In the end, it's all about delivering for patients. For me, the most satisfying memories are of those difficult cases where we were not confident we could save the eye but we did, and for the patient it was a miracle. In our collaboration with PSI, everyone went the extra mile. If a machine didn't function, then a technician would crawl into a tunnel during the night to make sure that a patient from Italy could be treated the next day. It's the opposite of bureaucracy. For me that is very motivating."

Details of current procedures for patients receiving treatment on the gantries and at the OPTIS2 facility can be found on the Center for Proton Therapy website (www.protontherapy.ch).

Plans for a third gantry
According to Damien Weber, 2010 was the pivotal year, when the balance tipped and it was clear that the future of proton therapy lay in scanning, not scattering. The technologies developed at PSI were being replicated around the world.

"In 2009 the Rinecker Proton Therapy Center opened in Munich, using the same type of cyclotron that we use, and using pencil beam scanning on a gantry engineered by the same company that we collaborated with," says Damien Weber. *"It was not a significant victory for pencil beam scanning in itself, but soon afterwards the first pencil beam scanning systems came into operation in America – at MD Anderson in Houston, at the Massachusetts General Hospital in Boston and the Roberts Proton Therapy Center in Philadelphia. When the Americans started to have pencil beam scanning there was a sudden change of paradigm. People weren't thinking the Swiss were mad any more: they were acknowledging that pencil beam scanning had a future. Now literally every proton therapy centre has pencil beam scanning."*

It was a time of remarkable development at PSI too. The new OPTIS facility was complete. There was a full programme of patient treatment using Gantry 1. The new state-of-the-art gantry was being installed. And by 2010, another significant project had been added to the Center for Proton Therapy's already considerable workstream: a third treatment gantry. This time, the project was not at PSI's own instigation. Instead, it was a mark of growing national interest in the potential of proton therapy facilities to serve the health needs of a population.

The new gantry project was first mooted at a meeting in January 2009 when representatives from the Swiss Canton of Zürich proposed a collaboration between the

University Hospital of Zürich and PSI to build a third gantry to provide proton therapy for the people of Zürich.

"We had already received 20 million Swiss francs from the Swiss Canton of Aargau for Gantry 2, and then our neighbouring canton, Zürich, became interested in proton therapy," says David Meer. *"They made studies looking into whether they should build their own proton facilities, and decided that a better option was to invest 20 million at PSI to build a third gantry."*

In 2010 PSI signed an agreement with the University Hospital of Zürich and the Canton of Zürich: PSI would construct a third gantry for clinical research and for patient treatment. An application for financing funds was made to Swisslos-Fonds, the Swiss Lottery Fund. By 2012 the Parliament of the Canton of Zürich had approved the 20 million Swiss francs required, and options for the layout and location of Gantry 3 within the PSI proton therapy facility were proposed.

A key question had to be decided. Who should build the gantry? Such was the influence of Gantry 1 and Pedroni's innovative scanning technology, that there were now options beyond PSI building its own gantry. Commercial companies were manufacturing proton therapy systems for hospitals around the world. Should they build one for PSI too?

David Meer: *"At the time, Gantry 2 was still not fully running and there were some pessimistic voices, worried whether it would ever work, saying that a safer approach was to go to a commercial vendor because there were now five or six of them offering pencil beam scanning systems. The other option was to produce a replica version of Gantry 2. It was a time when we had to find our niche, because others were now doing the same thing as us. It was not an easy decision."*

Martin Jermann remembers the long discussions at PSI. *"Even today, some people are not happy about the decision we eventually took. Gantry 2 had been mostly built by PSI, using our own engineers – with only the main mechanical structure being constructed by an outside engineering company. But now we had no time to construct a gantry, a copy of Gantry 2. As a research institute, it was not our job to make copies, so it seemed certain we would outsource the construction to a company with knowledge and experience in the field."*

Changing times

The four years between the first discussions about Gantry 3 and a decision on how to proceed was a period of significant change and some self-examination for the proton therapy team at PSI. Further advances in treatment delivery were accompanied by significant staff changes and discussions about the way forward – partly raised by the issue of whether to contract out work on the new gantry.

The Center for Proton Therapy now stood as a respected clinical facility within a research institute. So to what extent should research and development still play a part? And now that PSI had demonstrated the value of proton therapy for some hard-to-treat tumours, should it now start to investigate its value for other, more common cancers? What was the role of proton therapy in the fight against cancer both nationally and globally?

In the 2010 PSI Scientific Report, Eugen Hug set down an agenda. "We now consider the present cancer indications as established, and have applied to the Swiss Federal Office of Public Health for permanent approval of proton therapy," he wrote. "The major technological emphasis of the new-generation spot-scanning system will be its application for moving targets. Appropriate protocols for liver and lung cancer are under consideration." Hug also pointed to the prospect of evaluating scanning-based proton therapy for common cancers such as those of the breast, prostate and lung (the focus was later to fall on lung cancer, where the specific advantages of proton therapy were becoming clear).

A year later Gudrun Goitein, who had played such a major part in establishing the clinical efficacy of pencil beam scanning, developed on Hug's report, summing up in the scientific report what had been learned about proton therapy at PSI and setting down her convictions about the future.

"Protons offer high precision in dose deposition," she wrote. "The medical question is where this precision is wanted or needed. Optimized spatial dose conformation can allow increased target doses without increasing damage to healthy tissues and organs. This concept is useful to treat relatively radiation-resistant tumours. Overcoming inhomogeneity and imprecision in dose deposition caused by organ and/or target motion can be avoided by really fast (re-)scanning, which then allows for efficient irradiation of, for example, lung tumours or lesions in the mobile parts of the abdomen and pelvis.

"Reduction of unnecessary radiation dose to sensitive anatomical structures or compartments (e.g. the brain, optic nerves, spinal cord, kidneys, etc) is the general aim in any form of radiotherapy. Paediatric treatment is the most demanding in this regard..."

"The decision of the Canton of Zürich to invest in a third proton gantry at PSI reflects the belief that making better use of protons in modern cancer therapy requires extension of the indications, mainly towards frequent diseases, the conduct of clinical studies and research, all resulting in the need for higher treatment capacity."

Goitein had returned to head the Center for Proton Therapy ad interim on Eugen Hug's departure in the middle of 2011. She retired in 2013. Martin Jermann also retired, in 2012, along with Eros Pedroni – the physicist who, over 35 years, had led innovations that had changed the global landscape of radiation therapy.

Pedroni left, he says, proud of what had been achieved with Gantry 2, which was about to come into operation. The technological progress made during Pedroni's time at PSI could be measured by the time it took for patients to be treated with pions and protons. Back in the 1980s, pion therapy required 30-minute treatment sessions. Later, proton scanning sessions on Gantry 1 had taken three minutes. Now, on Gantry 2, sessions were to take just ten seconds. Fast scanning had reached its peak.

Eros Pedroni: *"We now knew it was feasible to paint a litre volume in less than ten seconds, so had proved the feasibility of treating moving targets within a single breath-hold, repeatedly per fraction. In my eyes, the concept of time-driven scans using beam intensity modulation could be a new milestone in the development of proton therapy. This was, for me, reason to retire in a good mood, knowing that Gantry 2 was potentially ready for redefining the future of proton therapy. It was now for the younger generation to demonstrate this potential, as had been the case with Gantry 1 and IMPT."*

David Meer, who succeeded Pedroni, regards him as a visionary. *"He had a great intuitive understanding of physics. He understood what is needed, what is important and what is not important, and could translate his ideas into a language easily understood by physicians. I really appreciated it that when he left, he said: 'Now it's time for the next generation. I had my chance to give you my advice; now it's for you to take over.' He has influenced me a great deal."*

A decision made

Damien Weber became head of the Center for Proton Therapy in 2013, succeeding Eugen Hug and Gudrun Goitein. One of his first actions was to urge the Swiss Federal Office of Public Health to draw up a strategy on proton therapy for the country – a continuing campaign that only bore fruit in 2020 (see Chapter 9). His other priority was to get patient treatment on Gantry 2 started as quickly as possible.

After a six-month shutdown of clinical work to enable the commissioning of Gantry 2, Weber was determined that there should be no more delays. Final checks proceeded swiftly and the first patient was treated on Gantry 2 in November 2013.

"The beam characteristics of Gantry 2 are a game changer in the proton therapy field, with beam sigma values of 2.2 mm for pencil beam scanning (PBS) at high energy," wrote Weber in a Center for Proton Therapy newsletter. "This Gantry will enable the CPT team to deliver PBS to cancer patients in an unmatched conformation, with rescanning and repainting capabilities in due time.

Figure 63. Physicists and engineers in the Gantry 2 control room, as the first patient is treated on the gantry in 2013. From left to right: Pablo Fernandez, Stefan Danuser, Francis Gagnon, Christian Bula, Martin Grossmann and Serena Psoroulas.

Figure 64. Delivery and installation of Gantry 3. One half of the main front ring structure of the gantry arrives at PSI and is then mounted onto the other half. The patient couch will be installed into this structure. Pictures: Paul Scherrer Institute

"The clinical operation of this treatment unit is a major milestone for the Center for Proton Therapy and puts PSI once more at the forefront of proton therapy innovation."

In the same year, the PSI directorate made the decision that Gantry 3 should be built by a commercial partner – but this would be a research collaboration rather than a direct purchase, with PSI drawing up the technical specifications. There simply wasn't time for PSI to build the gantry itself.

There were several manufacturers now marketing whole proton therapy systems. They had experience in certifying gantries as a medical product – an important requirement for PSI in view of new regulations. PSI defined the new gantry technical specifications based on those of Gantry 2 – ultra-fast energy selection was essential. After a careful evaluation of the offers, in 2014 PSI decided to contract Gantry 3 to Varian Medical Systems.

"We specified to Varian that we would like to have fast scanning, with layer energy change of maximum 200 milliseconds," says Martin Jermann. *"All the companies were used to doing this in seconds, so we said that if you want the contract, you will have to implement a 200-millisecond technology in your product and we will license this technology to your company. Varian was interested in doing that for its future devices."*

Varian started construction in 2014, while PSI prepared the infrastructure to accommodate Gantry 3. This involved extending the proton therapy facility building. In 2015, the Gantry 3 pit and beam line were constructed, the bulk of hardware had arrived and installation of components began. Technical commissioning of the treatment unit began in the autumn, with the first patients due to be treated in late 2016.

Figure 65. The extended Center for Proton Therapy in 2014. Picture: Paul Scherrer Institute

"A joint clinical programme will be established with the University Hospital of Zürich and both teams on either side of the river Aare are very excited in this endeavour," wrote Damien Weber in the August 2015 newsletter.

Another landmark had been reached in 2015: the 1000th patient had been treated on a gantry at PSI.

Pushing the boundaries of child treatment
More than one third of the patients now coming to the Center for Proton Therapy were children. With the reorganisation and extension of the proton therapy facility required by the construction of Gantry 3, facilities for child anaesthesia and recovery were also changing: the recovery room was moved next to the anaesthesia room to provide direct access, and adults and children had separate waiting areas.

Between July 2004 and December 2021, 390 young children were treated for cancer under deep sedation or anaesthesia. The most commonly treated child cancers were, and still are, brain tumours and sarcomas.

Over the years, technology and treatment approaches have evolved. Different combinations of sedating medicaments are now used to improve blood pressure stability. Vital signs-monitoring technology has been upgraded. And with growing experience in the management of small children under sedation, the specialists from Zürich Children's Hospital feel confident about treating increasingly ill and small children at PSI.

In 2014, they treated a child under the age of one for the first time: a ten-month-old boy weighing 7.8 kg, who had a brain tumour and had to be treated with a mask to fixate his head.

Then, in October 2016, the first paediatric patient was treated on Gantry 2 under anaesthesia. On Gantry 1, the only way to treat some tumours was with the child face down, but this made anaesthesia very difficult. On Gantry 2, it was possible to treat most kinds of tumours in the supine position, making sedation much simpler and safer. In 2018, the youngest ever patient was treated with scanned protons – just four months old and weighing only 5 kg.

"We were a bit challenged about how it would work with such a young child," says Martina Frei, *"but our advantage was that we had already learned a lot at the chil-*

dren's hospital about doing MRI scans of very young children under sedation. It was still a very special situation."

Other extremely ill children who presented significant treatment challenges began to be treated more often.

"There was a child with a tumour in her skull base. The tumour was very close to the respiratory centres and she also had a lot of respiratory tract infections all the time. Surgery of the tumour was too risky and her family had applied for treatment at PSI three times before. The first time doctors said no because she was too sick for anaesthesia. The second time, she was accepted but developed sepsis, so we had to postpone treatment. When we finally came to treat her in 2016, she had run out of all other options. When she went under sedation we were astonished: her respiratory rate was around five breaths per minute, which is as slow as a tortoise hibernating. We were sitting there, watching and thinking: 'Please... next breath, next breath.' This was really challenging.

"But she did great. She had no other problems during treatment, no complications over the weeks that followed. The last I heard she is now ten, at school, active and enjoying life.

"The intention is always to cure, but you never know. And we do have children who die a few months after treatment because they have had a very fast relapse. It happens, unfortunately. But on the other hand, we sometimes get to see the success stories long after. The first two child patients treated at PSI, in 2004, visited us a few years ago, ten years after they had been treated. They are both fine and doing well, and so that was absolutely wonderful."

First treatment on Gantry 3
Under the supervision of project manager Jürgen Duppich, construction, installation and testing of Gantry 3 continued through 2015 to 2017. Although the gantry was commissioned externally, there was still much work to do for PSI staff. *"The main challenge for us was to interface the system,"* says David Meer, who had now been appointed head of technology development at PSI.

Martin Grossmann summarised the "interface" challenge in the Center for Proton Therapy newsletter.

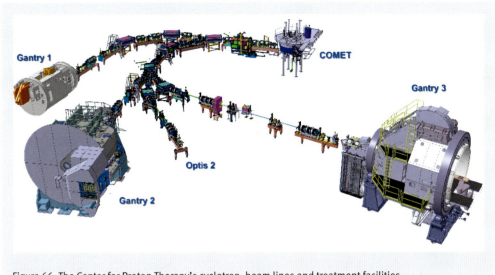

Figure 66. The Center for Proton Therapy's cyclotron, beam lines and treatment facilities.

"Gantry 3 is based on Varian's commercial ProBeam product. As such it comes fully equipped with its own systems to apply the proton beam in a precise and safe way. The situation of Gantry 3 is however different to ProBeam: in the standard system the whole facility is provided by Varian. But for Gantry 3 the accelerator and the beam line, along with their respective control and safety systems, have been developed by PSI or third parties. The challenge was to connect these worlds and the different technologies involved. The approach to solve this problem was to leave the existing systems mostly untouched and provide interfaces by newly developed interfaces called 'adapters'."

Martin Grossmann explains: *"These adapters are key to correct and safe patient treatment. They set the beam parameters according to the clinical demands, and if there is a problem, they quickly switch off the beam to prevent potential damage."*

The first patient was treated on Gantry 3 on 16 July 2018. In the August 2018 newsletter, Damien Weber paid tribute to Jürgen Duppich and Alexander Koschik for steering the Gantry 3 project, and Martin Grossmann and Christian Bula for their role in ensuring that the first patient was treated successfully and safely. Integrating an externally developed technology into the complex IT architecture of PSI was "not an easy task but one of paramount importance, as control and safety systems are the brain and 'safety net' of our medical devices delivering proton radiation to our cancer patients."

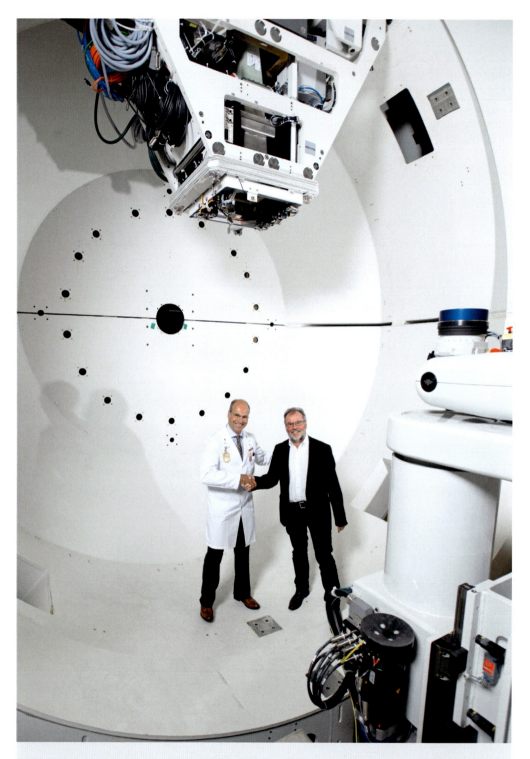

Figure 67. Gantry 3 Project Manager Jürgen Duppich (right) and Head of the Center for Proton Therapy Damien Weber pictured in Gantry 3 during its assembly. Note the nozzle above them and the robotic table front right

By now, PSI's long experience and expertise in building control systems for proton therapy facilities was unique. It wasn't only Damien Weber who was impressed with the skills of Martin Grossmann, Christian Bula and their team.

"One of the nicest compliments I ever received for my work came from one of our radiation technicians," says Martin Grossmann. *"She had heard that in several early proton therapy centres the development of the control system led to significant delays in the project schedule. She came to me and asked: 'Why is the control system a problem everywhere else, but not at PSI?'"*

With a total weight of 270 tons and a diameter of 10.5 metres, Gantry 3 was, and is, the largest machine installed to date at PSI's Center for Proton Therapy. On its launch, Damien Weber said: "With Gantry 3, we can offer highly effective proton therapy to more patients than ever before."

Patient treatments on Gantry 1 ended in December 2018, although the gantry continues to be used for radiobiology experiments (see Chapter 9).

Clinics and research: a successful balancing act
Alongside the development of new treatment facilities, the Center for Proton Therapy has developed an academic research and development programme in collaboration with Swiss universities.

Since 2007 it has had an affiliation with the Physics Department of the Swiss Federal Institute of Technology (ETH), Zürich, through the appointment of Tony Lomax as a professor of medical physics. Since 2012 there have also been affiliations to the medical schools at Zürich and Bern universities through the appointment of Damien Weber as a professor at both institutes.

This has led to a highly successful graduate educational programme, with more than 25 PhD students from the Center for Proton Therapy having graduated from these institutes by the end of 2021. Research topics include advanced delivery methods for more efficient and effective treatments, and novel workflows for quickly reacting to anatomical changes during treatment (adaptive therapy).

Also being researched are so-called 4D treatments for the treatment of tumours in the thorax and abdomen. Many tumours in these regions are literally "moving targets" because of the effects of breathing. The Center for Proton Therapy has conducted many pioneering studies on approaches to treating tumours in these areas, from repainting and "gating" (delivering only within a limited part of the breathing cycle), through to a technique called "tumour tacking", in which beam movement is synchronised with the tumour motion.

Another important part of current research is the detailed analysis of patient outcomes in order to understand what aspects of the treatment, and which clinical features of the patient, may affect treatment outcomes or side effects.

"The more we understand these aspects, the more we can feed back into the research programme, thus making even greater improvements to the efficacy of proton therapy," says Tony Lomax.

The global success of pencil beam scanning
Today PSI treats around 160 patients a year on Gantries 2 and 3: a significant increase since the pioneering days of Gantry 1 in the late 1990s, when just three dozen or so patients benefited annually. Since 1996, more than 2,000 patients from all over Europe have been treated on the PSI gantries, of whom more than 700 have been children or adolescents.

But the success story of the PSI gantries goes well beyond the number of patients treated – or the number of lives saved. The pencil beam scanning that was pioneered on Gantry 1 and then perfected on Gantry 2 has served as the model for proton therapy facilities around the world, and fast scanning has become the leading method of proton delivery.

In the past decade, a growing body of evidence has supported its efficacy, particularly for treating children and tumours in the brain and spinal cord. Notable among the research is a paper by Damien Weber and colleagues, published in *Radiotherapy and Oncology* in 2016, on the long-term outcomes of patients who had received pencil beam scanning proton therapy for skull-base tumours at PSI.

With its long history of providing pencil beam scanning with protons, and records going back to 1998, PSI was able to examine the outcomes of 222 patients, at a mean follow-up period of over four years. It found that long-term local tumour control was achieved in more than 70 per cent of patients, with the long-term toxicity-free survival rate being 87 per cent.[22]

With the evidence base growing, and the needs of potential beneficiaries (particularly children) urgent, the global rise of proton therapy in the past decade has been remarkable. There are today around 100 proton therapy units in clinical operation around the world and 85 per cent of them are using pencil beam scanning technology.

For some of those who worked on the remarkable cutting-edge technology of Gantry 2 at PSI, there is a sense of sadness that, though Gantry 2 inspired the proton therapy boom, plans for PSI to commercially market and sell replicas never bore fruit. The only other Gantry 2 system in the world was licensed to the MedAustron particle therapy facility in Wiener Neustadt, Austria, but it is not yet in clinical operation.

"You can have the most beautiful scientific ideas, and then produce them in the best way you possibly can, as we did," says Martin Jermann. *"But then industry, the hurdles of certification and the market decide. When I first came into this clinical field of radiation oncology, I was convinced that the best will always win, because it's used to cure cancer patients. But sadly, that is not the case."*

Today, the Center for Proton Therapy at PSI can no longer claim that, in terms of clinical innovation in radiation oncology, it stands way ahead of any other particle therapy centre in the world.

"I remember Martin Jermann telling me in 2007 that we were still ten years ahead of anyone else," says David Meer. *"In 2013 we were maybe two to three years ahead. Today, there are so many competitors and we are all more or less on the same level. It's not that research and development aren't still supported at PSI. They very much are. It's just that, nowadays, it takes more effort and money to move one step ahead than it ever did before. Twenty years ago, you could have a big effect. Nowadays, you have to invest more and more."*

The issue of cost has continued to pursue proton therapy as its use has spread around the world. Whether, realistically, its applications can or should spread

---

[22] D. C. Weber, R. Malyapa, F. Albertini, A. Bolsi, U. Kliebsch, M. Walser, A. Pica, C. Combescure, A. J. Lomax, R. Schneider, 'Long term outcomes of patients with skull-base low-grade chondrosarcoma and chordoma patients treated with pencil beam scanning proton therapy', *Radiotherapy and Oncology,* 120 (2016), 169–74. https://www.sciencedirect.com/science/article/pii/S0167814016311148?via%3Dihub.

beyond those which PSI has demonstrated as effective over more than a quarter of a century is open to question. Eros Pedroni, the man who played such a significant part in developing proton therapy as we know it today, is still unsure whether the treatment can become firmly established as part of global health-care systems.

"The problem with proton therapy is that you need big accelerators, and that makes things costly. Once it's working, the operating costs are not so high, and it will last for 20 years or so, but you need a very big investment at the beginning and people are very reluctant to risk their necks on a big project. Because it's such big machinery, proton therapy gantries need to be very centralised so that all hospitals can send their patients there. If you have a drug you can send it to any doctor, but at the moment proton therapy is so big you have to provide it from one place for the whole country."

The hope for proton therapy, believes Pedroni, lies in what has been learned in the past 40 years at PSI, where the spirit of free innovation, teamwork and scientific commitment were encouraged and allowed to thrive.

"Proton therapy is a difficult interdisciplinary endeavour," he says. *"If the spirit of collaboration, mutual understanding and respect across the various disciplines – medicine, physics, engineering, industry, hospitals, authorities, financing – is not strong, failure is not far away."*

Damien Weber, currently steering PSI's Center for Proton Therapy through the realities of 21st-century financing and regulation, agrees. The environment and values of the Paul Scherrer Institute have always been key to the development of proton therapy, and will continue to put it in the first rank among global particle therapy facilities.

"I still feel the wow factor about proton therapy that I had when I first came to PSI, and I have huge expectations about its potential," he says. *"I still value what we have at PSI as extraordinary. We are the only place, in Switzerland at least, where you have a research facility with the drive you find in American campuses. You have expert clinicians alongside every type of scientist, who can bring substantial know-how to what you do. There's always someone to bring another perspective. Everything is possible."*

Chapter 9

The future of proton therapy at PSI

By 2021, around 250,000 people had been treated with proton therapy globally. Nearly all new centres use pencil beam scanning and modulation techniques based on PSI's pioneering work. The debate about scanning versus scattering is over.

To quote the proton therapy entry in Wikipedia, that online gauge of popular understanding: "The newest form of proton therapy, pencil beam scanning, delivers therapy by sweeping a proton beam laterally over the target so that it delivers the required dose while closely conforming to [the] shape of the targeted tumor. Prior to the use of pencil beam scanning, oncologists used a scattering method to direct a wide beam toward the tumor.

"Delivery of protons via pencil beam scanning, which has been in use since 1996 at the Paul Scherrer Institute, allows for the most precise type of proton delivery known as intensity modulated proton therapy (IMPT)." [23]

Demand for proton therapy based on pencil beam scanning has grown over the past decade as global awareness has increased. Its potential to save the lives of children with previously untreatable cancers has, in particular, drawn publicity. Some parents, unable to access proton therapy in their own country, have staged protests and started fundraising campaigns to finance trips to access proton therapy privately abroad. Responding to the evidence about the value of proton therapy for children, many countries have made significant investment in new facilities.

"We are now experiencing an exponential increase in proton therapy centres around the world," says Damien Weber. *"Japan has a large number of facilities, and if you break down the number of gantries per million of population, Japan is*

---

[23] 'Proton therapy', Wikipedia, https://en.wikipedia.org/w/index.php?title=Proton_therapy&oldid=1089332586, accessed 20 June 2022.

way above all other countries. In the 2010s there was a significant increase in proton therapy centres in the US, and we are currently seeing a similar increase in Europe. In the UK, there are now two NHS facilities and at least two or three functional private facilities. It is quite extraordinary that a country that had no proton therapy centres in 2018 now has four or more."

But what of the future of proton therapy? Will investment prove justified? Will demand continue? And where does PSI stand in the national and international picture? Definitive answers are difficult. The potential of proton therapy – and its future development into new and even more effective radiotherapy techniques – continues to excite scientists and clinicians alike. But what happens next depends, to a large extent, on the economic and political realities of modern health care.

Damien Weber: *"As with other new technologies in health care, there has been a massive increase in interest – some might call this hype. We are now approaching a plateau. What happens next could go either way.*

"We are now in a very different situation than we were in the 1980s, when you had physics-dedicated centres delivering these innovative treatments to very selected patients. There are now many health institutions delivering proton therapy as part of cancer care, but there is uneven access. For instance, in the UK, you have around one gantry per 10 million people, whereas in the Netherlands you have a gantry for every 2.5 million people. Either the British are right or the Dutch are right, but whichever way you look at it, access to proton therapy is very variable in Europe."

The global picture of proton therapy costs and benefits
Work assessing the effectiveness and cost-effectiveness of proton therapy for different patient groups is ongoing. There is no question that proton therapy is of proven benefit when it comes to the treatment of children: their susceptibility to the side effects of radiation therapy, the risks of long-term toxicity and the potential costs of future care and screening make the expense of proton therapy economically and morally worthwhile. Today, approximately 40 per cent of patients treated at PSI are children.

The evidence is also good for the main cancers that PSI currently treats: brain and spinal tumours, sarcomas and head and neck cancers. The evidence base is still building – after all, modern proton therapy techniques have only been widely used for a decade. There are now global research efforts to answer the question: does proton therapy show sufficiently significant benefits over conventional radiotherapy to justify its extra cost?

For example, the University Medical Center Groningen has a growing database of head and neck cancer patients treated with both conventional radiotherapy and proton therapy, which will allow high-quality comparisons of outcomes. Head and neck cancers are the fourth most common cancer treated with radiation therapy, after lung cancer, prostate cancer and breast cancer. Already results are indicating that, for some types of head and neck cancers, around 95 per cent of patients benefit more from having proton therapy than conventional radiotherapy.

Meanwhile, the European Particle Therapy Network, a task force of Europe's professional body for radiation oncology, ESTRO, is beginning a trial to compare proton therapy with conventional radiotherapy for gastrointestinal tumours.

Damien Weber: *"We need this kind of comparative research, because proton therapy involves a very complicated piece of technology which requires a big investment, and it also often involves considerable additional cost factors for patients. We have to show that it is worth it."*

Future priorities at PSI
The Center for Proton Therapy (CPT) at the Paul Scherrer Institute is influenced by this global context but also influences it through its research and innovation. The centre is firmly established on a national and international level, continuing to build on its history of clinical excellence combined with scientific innovation and research. There are currently three main priorities: improving access to proton therapy; improving treatment precision and efficiency; improving knowledge of tumour and normal tissue responses to proton therapy.

"CPT's mission is to provide optimal and individualized treatments for Swiss cancer patients... to generate high-quality clinical data and to participate in high-risk research development that in turn will maintain PSI's position as a world-leading proton therapy centre," reads its strategic plan for 2021 to 2024.

As has always been the case at PSI, the motivation to do something new or better is strong. Although proton therapy is already a highly precise treatment technique, finding ways in which precision can be further improved is a cornerstone of the CPT's research and development strategy. The centre continues to develop advanced

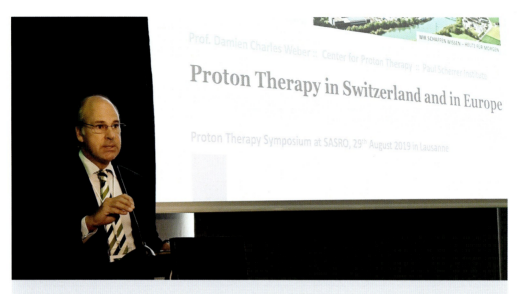

Figure 68. Damien Weber addresses the Proton Therapy Symposium at the annual meeting of the Scientific Association of Swiss Radiation Oncology (SASRO) in Lausanne in 2019.
Picture: Peter Brandenberger www.kongress-foto.ch

methods for mitigating the motion of patient organs and targets during treatment. It is also developing ways of re-optimising treatment plans on a daily basis and investigating the potential of ultra-fast, high-dose-rate treatment techniques.

Tony Lomax, who is today chief medical physicist at PSI, feels the need to innovate in radiation therapy still – and believes the culture of PSI inspires it.

"You're always pushing forward," he says. *"We want to do new things. This is why we started treating lung patients in late 2020. Treating the moving lung is very complicated, but we've done a lot of research on it. Every time you make these kinds of developments and start new types of treatment, there's an element of excitement but also an element of worry, because although we've got a complex delivery system with very effective safety systems, we have an even more complex target, which is the patient. There are no eureka moments. There's no point at which you say: 'That's it, we've done it.' It's a continuum. But that's okay. It's what keeps us all motivated, and keeps us all doing what we're doing."*

Innovation in organ motion and lung cancer
The aspiration set down by Eugen Hug in 2010 that CPT should begin to evaluate the effectiveness of pencil beam scanning for the most common cancers has today become focused on lung cancer – the most common cause of cancer death in the

world. PSI's experience in researching how to accurately irradiate a moving target with protons is particularly important here: clearly the lungs move as long as a patient is breathing.

In 2020 the Federal Office of Public Health in Switzerland accepted non-small cell lung cancer as a new indication for proton therapy treatment within the framework of a five-year prospective international trial assessing the clinical benefit of protons over photons. This means that the treatment can be reimbursed by Swiss health-care insurance. Current PSI research in lung cancer includes investigating different breath-hold techniques, new ways of keeping the lung as stationary as possible during irradiation and ways of making the proton beam accurately follow lung movement.

The beauty of the ultra-fast pencil beam scanning achievable at PSI is that it can minimise the impact of lung movement on effective treatment: rapid rescanning and breath-holding to keep the lung stationary are perfectly achievable.

"There's a lot of potential for treating lung cancer here, because we can control moving targets using the scanning and rescanning techniques that Eros developed," says Martin Jermann. *"Other centres have already found that they can treat some cases of lung cancer with scanning because the tumour is in a part of the lung that is not moving so much – perhaps 5 mm, as opposed to larger movements in other lung segments.*

"Now we are looking to be able to treat in all parts of the lung, and the development could go worldwide. The Gantry 2 technology is so fast that we can treat a half-litre volume in less than one minute."

Ultra-high dose rate (FLASH) proton beam delivery
Another area that CPT is researching is FLASH – a completely new type of radiotherapy based around delivering extremely high doses 400 times more rapidly than conventional radiotherapy. It involves doses of up to 1,000 Gy per second, roughly 100 times higher than conventional treatments. If proven, it could spare patients many weeks of radiation therapy and significantly reduce side effects.

Early experiments in animals have indicated that the high dose rates of FLASH actually minimise radiation damage in healthy tissue while neutralising tumours as effectively as other forms of radiation. Different types of irradiation system can be adapted to produce FLASH radiotherapy dose rates – including electron linear accelerators and proton pencil beam scanning gantries.

In collaboration with the Centre Hospitalier Universitaire Vaudois (CHUV) in Lausanne, the CPT at PSI is investigating how effective FLASH might be using protons. The research utilises Gantry 1, which is now closed to patients but plays a vital role in experimentation. After some adaptation, Gantry 1 is now capable of delivering 9,000 Gy per second.

Damien Weber: *"FLASH is not a new idea. Its origins lie in research from the 1950s and 1960s. The research endeavour was stopped as a result of technological issues and dubious results. Quite recently, the idea of ultra-high-dose rate delivery was investigated by a joint team from CHUV and the Center for Proton Therapy at PSI. They found that if you deliver very high dose rates of radiation to lung tumours, the surrounding tissue is spared. It seems completely counter-intuitive, but it seems that the high dose rates actually deplete the surrounding tissues of oxygen, in the process making them more resistant to radiation."*

Experiments at CPT are studying this oxygen-depletion effect using molecular imaging, measuring the metabolites of oxygen in exposed tissue.

"Using protons as opposed to other particles might give the opportunity to enhance the benefits of FLASH, given our ability to both accurately control dose deposition and conform the dose using pencil beam scanning. Combining these well-known advantages of proton beams with their capability of delivering ultra-high dose rates may hasten the clinical translation of these innovative technologies in the near future.

"We are a long way from clinical implementation, but it could be a very interesting modality. If FLASH works, we will have to reinvent radiotherapy. Up until now, we've been mostly delivering radiotherapy at 1.8 to 2 Gy per fraction, over five to seven weeks. FLASH would mean a couple of fractions over one week, done."

Challenges and collaborations
In terms of patient care, PSI's place at the forefront of proton therapy within European and Swiss health care is well established. But there are challenges. Does PSI have the potential to deliver more, and to more patients? The future of proton therapy at PSI is intrinsically linked to the future of proton therapy in Switzerland.

Activity and ambition at PSI's proton therapy centre has always been to some extent defined by its relatively remote location. It is not as accessible to patients or health-care professionals as a large city-based hospital. This has meant that collaboration with hospitals has always been – and continues to be – fundamental to its success.

"Being in the middle of the countryside has its advantages, but in terms of managing cancer patients who need high-tech treatment and multidisciplinary teams, it's always a challenge," says Damien Weber.

"We have to make sure we meet all patient requirements for top-quality care, and the way we have reacted to this is through clinical collaborations with hospitals. Our engagement with the Children's Hospital in Zürich – the largest in Switzerland, which performs clinical research alongside excellent patient care – has been extremely successful.

"Since I took over as head of the CPT in 2013 I have initiated a number of other clinical collaborations. One is with the large cantonal hospital of Baden, with whom we have an active clinical collaboration in diagnostic radiology. They use our new MRI scanner, and are involved in the interpretation of all our radiological studies. This has been proven to be an extremely successful collaboration because we have input from specialist radiologists who can help us define target volumes for treatment and provide expert opinion and patient evaluation.

"Additionally we have a strong partnership with the cantonal hospital of Aarau. Our clinical trial on lung cancer is being run in association with the hospital.

"We have another clinical collaboration with the University Hospital of Bern. We jointly fund a radiation oncologist, encouraging patient referral from the hospital in Bern and performing joint clinical research with the university hospital.

"We also have a clinical collaboration with the radiation oncology department at the University Hospital of Zürich, which was triggered by the Canton of Zürich funding Gantry 3. Zürich is one of our largest providers of child patients. In addition, we have treated in a collaborative setting a number of patients coming from St Gallen, Basel, Bellinzona, Lausanne and Geneva.

"These kinds of clinical collaborations are of paramount importance if we want to deliver optimal cancer care in this location."

Patient throughput

Today, PSI treats more patients annually than it did in 2010, but the increase has been modest. Some of this is accounted for by a decline in international referrals. Whereas in 2016, 37 per cent of patients came to PSI from outside Switzerland, today just 15 per cent are non-Swiss. Damien Weber says this was a necessary shift.

"It is right that the priority of a Swiss facility, supported by Swiss public funding, should be treating Swiss patients or those living in Switzerland regardless of nationality. But also treating international patients is a challenge. It is difficult to dovetail our treatment with the care of hospitals in the local country, and patients can become isolated and in need of intensive support in Switzerland if there is any kind of problem. It can also be very expensive and demanding for the patient to travel and stay in Switzerland."

The fact that Swiss patient numbers overall have not increased in the way foreseen a decade ago is partly because randomised clinical trials comparing proton therapy with conventional photon therapy are hard to perform, and don't attract as much funding as drug trials. Without the hard evidence of effectiveness that such trials bring, it is hard to extend the list of clinical indications for which proton therapy is officially approved.

The situation is also related to Swiss referral patterns. There is no lack of need for proton therapy. The problem is that patients are not being identified as likely to benefit from proton therapy and there are no incentives for hospitals to make referrals.

Damien Weber: *"The system works in such a way that if a hospital refers a patient to us, it loses income. So there is actually no incentive for radiation oncology departments to refer patients to PSI rather than treating them themselves with conventional radiotherapy. Because of that, we do not treat enough cancer patients who would benefit from proton therapy.*

"Our plan had been that by 2016 we would annually treat 600 to 800 patients who would benefit from proton therapy. But we are treating 350 to 400 a year."

This situation is not helped by the fact that the formulae being used by national health planners to estimate how much money should be allocated to proton therapy underestimate the clinical need. Again, this provides no incentive for referring to facilities like PSI, says Damien Weber.

"Business plans in some countries are being made on the basis that 5 per cent or so of patients who need radiotherapy would benefit from proton therapy, but this percentage is not based on any science – it's just been copied and pasted from country to country."

This widely used target percentage may be inexplicably low, but it is still higher than the current situation in Switzerland. The Center for Proton Therapy at the Paul Scherrer Institute is still the only proton therapy centre in the country, and captures just a tiny proportion of all radiotherapy patients.

Recent years have seen a number of plans to build further Swiss proton therapy facilities – both private projects and those in university hospitals supported by individual Swiss cantons. But the university hospital proposals have been refused permission by the Swiss Federal Office of Public Health and the rest have yet to take shape.

A new strategy
Switzerland, believes Damien Weber, needs to join the majority of European countries and have a national strategy for proton therapy – a plan that would allow services to be built around the needs of the population, and would give hospitals and centres such as PSI the opportunity to plan for the future.

"We decided to construct Gantry 3 in 2013 and treated the first patient in July 2018. Because of the infrastructure involved, what we do requires a huge time lead. If you don't have any strategy, how are university hospitals and private clinics going to be able to move forward in proton therapy? The health-care providers and the health system need to be able to foresee what will be needed – not tomorrow or the year after, but in five or ten years' time.

"There are three possibilities. One is that things stay as they are. The second would be to keep PSI as the centre for the German-speaking part of Switzerland, and build another for the French or Italian part. The third option would be to have multiple proton therapy centres, mainly in university centres, but also private initiatives.

The answer depends not only on Swiss politics but also on the general acceptability of proton therapy in Europe and elsewhere."

As far as the Center for Proton Therapy at PSI is concerned, Damien Weber would like to see it being part of a wider network in which every patient – and particularly every child – who needs proton therapy will have facilities available close to home. But whichever way the strategy goes, its mere existence will help the CPT at PSI define its role and priorities.

"I have always been a strong advocate for at least one other completely hospital-based facility, and if that happened, one way for PSI to go might be to reduce its treatment programme and focus more on clinical research, maybe finding ways to improve and develop pencil beam scanning. There's a lot of technical development to be done that could benefit proton therapy generally."

Conducting research that refines, improves and maybe revolutionises particle therapy for people with cancer will continue to be a priority for the Center for Proton Therapy at PSI. Those who work there are proud of its 40-year legacy of pushing boundaries, pursuing new ideas in the face of scepticism, testing them to extremes and sometimes, in the end, delivering something that really matters.

That pride is not just shared between PSI staff. Amid all his other work, David Meer prizes the opportunity to tell the public about what happens at the Center for Proton Therapy at PSI and why it is important.

"I've been here for around 17 years, and many things have been important to me. Being part of a team that put together Gantry 2 was possibly the most significant, but the public tours explaining the facility are also very important to me. There are around 4,000 visitors who come to the Center for Proton Therapy on guided tours each year. This communication with the public about proton therapy is a priority for PSI."

This book for the public is also a manifestation of that pride, and the value PSI places on spreading the word about the role of basic physics research and how it can be translated into applications that serve humanity – none more important than saving lives.

A patient story

What is it like to receive treatment for cancer at the Center for Proton Therapy at PSI? This is a first-hand account from a recent patient who was referred to the centre after previous attempts to treat a brain tumour had reached a dead end.

"After I had surgery on a brain tumour behind my left eye, there was good news and bad news. The good news was the depression symptoms it had caused were gone immediately. But the bad was that a remnant of the tumour was still in there. They couldn't operate again and the tumour was wrapped around a blood vessel. Chemotherapy would have destroyed much more than the tumour and conventional X-ray irradiation could have hit the optic nerve. So my doctors suggested I should go to the Paul Scherrer Institute for proton irradiation.

"I agreed right away. People were coming from other countries, far and wide, and I was glad that I got in. In the introduction at PSI, everything was explained to me in detail: how the therapy works, how long it takes and the risks.

"Then came two days of preparation. It was mostly about making a bite block and a cushion that fitted my head exactly so that the beams would hit the tumour with millimetre accuracy. Positioning is a very important thing in this therapy. You really have to lie correctly or the radiation will hit off target. That is checked again and again.

"When you arrive at PSI every morning for treatment, it doesn't feel at all like coming into a hospital. The therapy for my tumour consisted of 28 irradiations of 40 minutes each, five times per week over a period of six weeks.

"The therapy was absolutely pain-free. I was allowed to listen to music and the time passed quickly. I couldn't speak and you are alone in the room because of the radiation. But you are under constant observation with cameras. And if something comes up, you can just give a hand signal. Naturally the machine seems like a monster in the beginning – very imposing. I remember it made funny noises like a departing suburban train.

"The risks had been explained to me upfront. For example, I could have got headaches from the radiation or excessive fatigue. I was lucky, though. All I had was hair loss, but the hair grew back relatively quickly, after about three months.

"After the 28 treatment sessions, I didn't know whether the therapy would be effective or not. But with the follow-up exam three months later we found out. The tumour is still exactly the same size, but it's dead and it's not growing any more. That was a big relief. Additional check-ups followed every three months. They were always good and now it's half a year before I have to go back again.

"I had 100 per cent confidence in PSI because of the technology but also the people. They were very empathetic and professional, no matter who I was in contact with, from the secretaries to the radiology assistants and, of course, the doctors."

Index

A

Accelerator
13, 17-18, 21, 24, 27-30, 42, 45, 49, 84-85, 87, 93, 97-98, 106, 129, 134-137, 140, 142-143, 146-147, 165, 193, 204

Adaptive therapy
195

Animal
21, 45, 53, 78, 111-121, 123, 133-134, 203

B

Beam optics
38, 94, 108

Bending magnet
102, 151, 162-163, 177

Bite block
61, 69, 94, 154, 209

Blaser, Jean-Pierre
14, 21, 24-25, 27-31, 34, 42, 47-48, 55, 57-58, 66, 82, 83, 90-91, 95, 103

Blattmann, Hans
5, 29, 33, 35, 37, 45-47, 55, 66, 81-84, 86, 88-90, 92-93, 95-96, 99, 102-103, 112, 116, 119-120, 124, 133, 150

Bodendoerfer, Gerd
49, 51

Böhringer, Terence
99, 104-105, 128-129

Blosser, Henry
143

Bragg peak
16-18, 26-27, 39, 69, 87, 104, 139, 173

Bula, Christian
162, 176, 188, 193, 195

C

Children's Hospital Zürich
154, 157, 191, 205

Clinical research
34, 148, 180, 185, 205, 208

Collimator
38, 65, 70-71, 81, 93, 100, 165, 172-173, 183

COMET
22, 30, 146-147, 160-162, 166

Control system
13, 19, 123, 129, 146, 152, 160, 162, 164, 176, 195

Cooling system
97

Coray, Dölf
89-90, 99, 103-104, 122, 129, 131, 164, 169

Crawford, John
36, 47

D

Daum, Manfred
86

Deep-seated tumour
20, 29, 44, 51, 70, 83, 85-86, 91, 105, 127, 130, 150, 167-168, 175

Deflector plate
147, 151

Degrader
85, 87, 101, 142, 151, 161

Dosimetry
29, 32, 45-46, 62, 65, 69, 113, 129, 160, 164

Double-scattering
166, 172, 179

Duppich, Jürgen
146, 192-194

E

Eberle, Meinrad
14, 21, 103, 121, 124, 126, 134, 136-138, 142

Egger, Emmanuel
5, 61-62, 67-69, 71-73, 96, 115-116, 122, 124, 128, 131, 166, 182

Electron
15, 17, 25-28, 33, 39, 204

Enge, Harald
94

Eye tumour
7, 20-21, 54-55, 57-59, 61, 64, 67, 73, 75, 85, 113, 181, 183

EYEPLAN
59-60, 69

F

Fermi, Enrico
26, 33

Feurer, Rita
5, 154

FLASH
22, 203-204

Frei, Martina
5, 154, 157-158, 169, 191

Fritz-Niggli, Hedi
29, 34

G

Gailloud, Claude
54, 57-58, 73

Gating
196

George, David
163

Gillette, Ed
112, 118-119

Goitein, Gudrun
5, 11, 13, 68, 78, 92-94, 96, 99, 102-103, 106, 108, 111, 113, 116, 119, 124-125, 128, 131, 133, 135, 137-139, 141, 143-144, 146, 148-150, 152-154, 164-166, 168-169, 171, 175-176, 186-188

Goitein, Michael
59-60, 69, 92, 121, 136, 148

Gottschalk, Bernhard
141

Gragoudas, Evangelos
58-60

Greiner, Richard
5, 53, 66, 68, 75-81, 84-85, 88, 91-92

Grossmann, Martin
5, 13, 106, 120-121, 123-125, 128-129, 131, 162, 166, 188, 192-193, 195

H

Hilbes, Christian
152, 160, 162, 164

Hirt, Wilfred
55, 90, 103

Horizontal beam line
21, 94, 96-97, 147, 166

Hug, Eugen
166-167, 175, 179, 186-188, 202

I

Injector cyclotron
28, 30, 64, 66, 85, 136, 179-180

Intensity modulated proton therapy IMPT
12, 21, 87, 131-132, 138-140, 187, 199

Intensity modulated radiation therapy IMRT
130-132, 141

Ions
15, 17, 26-27, 82, 98, 106

J

Jaccard, Samuel
58, 65

Jermann, Martin
5, 11, 14, 21, 83-84, 89, 103, 121, 135-138, 141-142, 144-146, 161, 167-169, 171, 173-174, 179, 185, 187, 190, 197, 203

Jirousek, Ivo
44, 109

K

Kaser-Hotz, Barbara
5, 111-116, 118-120, 133-134

Kicker magnet
94, 104, 161

Kohout, Jaroslav
40

Koschik, Alexander
193

L

Larsson, Börje
112

Lin, Shixiong
99, 104, 128-129, 131

Linear energy transfer LET
16-17, 27, 39, 52, 82, 98

Lomax, Tony
5, 11, 99, 105-107, 124-125, 128, 130-133, 139-140, 148, 164, 166-167, 169, 195-197, 202

M

Mask
19, 61, 67, 182, 191

Mayor, Alex
162

Meer, David
5, 151-152, 160, 164, 176, 178-179, 185, 187, 192, 197, 208

Moving targets
147, 186, 187, 195, 203

Munkel, Gudrun see "Goitein Gudrun"

N

NA1 beam line
88, 94

NA1 bunker
88, 97

NA3 beam line
94-95, 97, 99-101, 104

Neutron
15, 17, 25-27, 43, 78, 84-85, 135

Nozzle
101-102, 104, 152, 165-166

O

Ocular tumour see "Eye tumour"

P

Paediatric anaesthesia
154, 156, 191

Passive scattering
86-87, 104, 128, 140-141, 149, 172-173, 175

Patient table
101-102, 107, 109, 134, 145, 151, 156-157, 177

Pedroni, Eros
5, 11, 26-27, 36-37, 39-41, 43, 45-49, 53, 55, 75, 80-83, 85-86, 88-89, 91-96, 98-99, 101-105, 107-110, 122, 124, 128, 130-131, 134-137, 140-141, 145-146, 148-149, 151-152, 163, 167, 169, 173-174, 176, 185, 187, 198

Perret, Charles
28, 33-34, 36, 39, 44, 47, 54-63, 65-69, 72-73, 180

Phillips, Mark
110

Pica, Alessia
180, 197

PROSCAN
7, 21, 110, 127, 137, 142, 144-147, 150, 153, 164, 171, 173, 178

PTCOG
10, 19, 90-91, 110, 116, 118, 125, 133, 148-149, 164, 173

R

Radiobiology
29, 33, 36, 45, 50, 77, 97-98, 119, 195

Range shifter
101-102, 104, 165, 166

Relative Biological Effect RBE
40, 46, 77, 78

Reist, Hans
147

Re-scanning
147, 186, 188, 203

Ring cyclotron
28, 30-32, 85, 98, 134, 160

Ring scanning
38-41, 43

Rutz, Hans-Peter
148, 169

S

Safety System
13, 44, 120, 123, 125, 135, 161-162, 164, 179, 193, 202

Salzmann, Miriam
37, 46-47

Scattering foil technique
86, 151

Schalenbourg, Ann
5, 180-183

Scheib, Stefan
99, 104-105, 130

Schippers, Marco
146, 147, 166

Slater, James M
91

Stereotactic chair
61-62, 166

Studer, Urs
51-52

Suit, Herman
81, 112

Superconducting magnet
30-31, 97

Superconducting spectrometer
29, 31-32, 35

Sweeper magnet
101-102

Swiss Cancer League
32, 47, 67, 78, 96, 104, 120, 168

Swiss Federal Institute of Technology ETH
2, 36, 41, 83, 152, 195

Swiss Federal Office of Public Health
32, 67, 141, 148, 179, 186, 188, 203, 207

Swiss Institute for Experimental Cancer Research ISREC
33

Swiss Institute for Nuclear Research SIN
21, 24-26, 28-34, 36, 38, 40-42, 44-45, 47-48, 51-55, 57-58, 60, 62, 64-69, 72, 75, 77-79, 81-82, 83, 85, 94, 97, 150

Swiss Institue for Reactor Research EIR
21, 24, 68, 83, 94

Swiss Lottery Fund
168, 185

Swiss National Science Foundation SNF
78, 82, 104, 120

Synchrotron
98

T

Tantalum clips
60, 67, 69, 183

Therapy Delivery System
123

Therapy plan
19, 123, 144

Therapy Verification System
123

Timmermann, Beate
148, 153-154, 169

Tissieres, Veronique
64

Tochner, Zelig
171

U

Ultra-high dose
203-204

University Hospital Zürich
105, 136, 141, 185, 191, 205

V

Vecsey, Georg
29-31, 33-34, 38-39, 83

Verwey, Jorn
165-166, 176

Volumetric repainting
176

von Essen, Carl
5, 26, 33-35, 37-38, 40, 42, 45-49, 51-53,
55, 58-59, 65-66, 75-76

W

Weber, Damien Charles
1, 5, 11-12, 139, 148-149, 153, 159, 173, 184,
188, 191, 193-202, 204-208, 217

Weiss, Markus
5, 154-155, 169

Y

Yukawa, Hideki
48

Z

Zografos, Leonidas
5, 54-55, 58-60, 62, 64-67, 69-70, 72-74,
166, 180, 183

A note about the photographs used

The photographs included in this book have come from a wide variety of sources, and the authors and PSI staff have done their best to identify original photographers so that credit can be given, and those pictured so that permission can be given. Given the historical nature of this book, this has not always been possible, and we apologise in advance if there has been any omission or error.

About the authors

Damien Charles Weber has been Head and Chairman of the Center for Proton Therapy at the Paul Scherrer Institute since 2013. He has currently a faculty position at both the University of Zürich and the University of Bern. He trained as a radiation oncologist at the University Hospital of Geneva and was a clinical and research fellow at the General Massachusetts Hospital/Harvard Medical School, after which he came to PSI.

His area of expertise is neuro-oncology and he has been Chair of both the Radiation Oncology Group of the European Organisation for the Research and Treatment of Cancer and the Scientific Association of Swiss Radiation Oncology. He is currently Co-Chairman of the European Particle Therapy Network (a task force of the European Society of Radiotherapy and Oncology, ESTRO) and the Publications Committee of the Particle Therapy Co-Operative Group. He is also a member of ESTRO's Scientific Council.

Simon Crompton is an award-winning journalist, writer and editor from the United Kingdom who specialises in health and science. He has contributed to and edited a range of publications for health and care professionals, was Medical Editor of the *Times* health section for five years, and contributed to *The Times* for more than 20 years. Throughout his career, he has worked as a communications consultant and writer for voluntary, professional and patient organisations across Europe. He has authored and contributed to several books and journal papers. www.simoncrompton.com